D0991154

Mechanical Failure Avoidance

Mechanical Failure Avoidance

Strategies and Techniques

Charles E. Witherell

McGraw-Hill, Inc.

New York San Francisco Washington, D.C. Auckland Bogotá
Caracas Lisbon London Madrid Mexico City Milan
Montreal New Delhi San Juan Singapore
Sydney Tokyo Toronto

Library of Congress Cataloging-in-Publication Data

Witherell, Charles E.
Mechanical failure avoidance : strategies and techniques / Charles E. Witherell.
p. cm.
ISBN 0-07-071170-4
1. System failures (Engineering) I. Title.
TA169.5.W57 1994
621.3—dc20 94-4635
CIP

Copyright © 1994 by McGraw-Hill, Inc. All rights reserved. Printed in the United States of America. Except as permitted under the United States Copyright Act of 1976, no part of this publication may be reproduced or distributed in any form or by any means, or stored in a data base or retrieval system, without the prior written permission of the publisher.

1 2 3 4 5 6 7 8 9 0 DOC/DOC 9 0 9 8 7 6 5 4

ISBN 0-07-071170-4

The sponsoring editor for this book was Robert W. Hauserman, the editing supervisor was Valerie L. Miller, and the production supervisor was Suzanne W. Babeuf. It was set in Century Schoolbook by McGraw-Hill's Professional Book Group composition unit.

Printed and bound by R. R. Donnelley & Sons Company.

This book is printed on recycled, acid-free paper containing a minimum of 50% recycled, de-inked fiber.

Information contained in this work has been obtained by McGraw-Hill, Inc., from sources believed to be reliable. However, neither McGraw-Hill nor its authors guarantees the accuracy or completeness of any information published herein, and neither McGraw-Hill nor its authors shall be responsible for any errors, omissions, or damages arising out of use of this information. This work is published with the understanding that McGraw-Hill and its authors are supplying information but are not attempting to render engineering or other professional services. If such services are required, the assistance of an appropriate professional should be sought.

Contents

Preface

An initial reaction from browsing through this book for the first time might be that engineered products, structures, and systems are in a bad way and that metals and alloys are unsuitable as materials for fabricating reliable mechanical components. Along the same lines, an impression of hospital visitors may be that human beings are an unhealthy lot.

Both observations would be incorrect, for both deal with abnormal groups—the exceptions. Most people enjoy good health and rarely require hospitalization. This situation is continually improving, as is evident from increasing life expectancy figures and decreasing infant mortality rates. Likewise, the vast majority of engineered components successfully fulfill their intended purpose. Despite increasing demands for higher efficiency, improved performance, and longer useful lives, engineered structures, machines, equipment, and products are, for the most part, more reliable today than ever before.

It is because of this—the conditioning to expect things to be reliable—that the occasional failure has become so newsworthy, shocking, and intolerable. Any failure that causes damage and injuries is one too many. But even engineering's worst critics must admit that the performance record and reliability of manufactured products and engineered systems have been steadily improving. Much of it has been the result of better understanding of materials and their response in the real world and conscientious and responsible management of technology.

Progress, however, requires more efficient use of energy and other resources, greater demands upon available materials, and intensified application of new and sometimes unproven technology. Accompanying all of this is greater risk. Jumbo oil tankers, jumbo jet aircraft, and jumbo manufacturing plants create jumbo risks and jumbo disasters when they fail. Higher-performance materials and more highly stressed and lighter-weight components operating under more demanding conditions, and for longer times, require more demanding and discriminating design and operating methods. We often get into trouble when we attempt to apply yesterday's design and manufacturing procedures to tomorrow's materials and performance requirements. It is a constant battle just to keep abreast of one small segment of technology let alone the multidisciplinary front that faces today's designers and production engineers.

Consequently, it is essential for those responsible for these kinds of activity—the engineers and managers—to understand the limitations of the materials they are working with and their susceptibilities to an array of degradation modes. And they must also realize that society, through its laws and regulations, has imposed additional design criteria, performance standards, and requirements to provide a safer workplace and living environment for its members.

The days of empirical engineering development, of trial and error, are long over. The marketplace is no longer the place to prove new products or materials or advanced technical concepts. Society may not demand perfection, but it does demand that engineered products and systems are not defective and do not cause harm, damage, and injuries.

The costs of failures to the manufacturer have risen exponentially, along with these demands. The double-barreled impact of law brought about by court decisions and legislated regulations has hit every industry. It is felt in defending lawsuits, product recalls, regulatory restrictions, increased insurance premiums, negative public image, eroded investor confidence, and shrinking profit margins. Failure avoidance in a manufacturer's products and its facilities is no longer optional but essential to the firm's very survival.

Strategies for failure avoidance can take many forms, as there are numerous failure mechanisms and virtually unlimited combinations of human deficiencies, component defects, and environmental conditions that can trigger failures, accidents, and damaging occurrences.

This book has four basic divisions for coping with this subject. The opening chapter considers origins and consequences of types of failures that are of major concern to mechanical engineers but that apply to a large extent to all engineering disciplines. A primary source of reliable guidance in avoiding failures is the study and analysis of prior incidents, and this is discussed in Chap. 2. Choice and implementation of strategies for failure avoidance require awareness and understanding of the causes of mechanical failures and familiarity with analytical tools for identification and priority classification of hazards. These are described in Chap. 3, along with techniques for minimizing failures and unscheduled downtime in operating plants and equipment. The last chapter covers day-to-day failure avoidance practices in design and production and how to keep flaws out of products and the manufacturing plant itself.

The scope of this subject is vast and embraces many engineering disciplines and scientific fields, although its focus is upon mechanical components. Therefore, all aspects cannot be accommodated in any great detail. Emphasis, however, is given to those failure modes and damage mechanisms that have been responsible for the majority of problems, including major catastrophes.

Most hazards and strategies mentioned could occupy a volume all their own and still not exhaust the subject. Therefore, each theme is supplemented with on-target references to assist the reader in locating additional information for implementing these practices and procedures. Some of the methods described are well established and their usefulness proven through years of successful use. Others have only recently been developed and have yet to be proven. Because of the rapid pace of development of these techniques in recent years in response to the growing need, developing trends in the field have been noted in an effort to address as wide a spectrum of readers' needs as possible.

It has become increasingly difficult in recent years to separate legal issues from engineering issues, and the increasing influence of court decisions and legislated regulations must be acknowledged. This has been done throughout the book wherever necessary and appropriate. However, because of diversities in the law among the many jurisdictions of this country and its changeable nature, especially in these fields, legal issues discussed here must be regarded as generally descriptive and illustrative only. Engineers and managers requiring interpretation and information on legal implications of specific situations and circumstances should consult legal counsel familiar with the details and the law of the particular jurisdiction and question at issue.

Similarly, failure avoidance strategies and procedures described here are for general information and education purposes and are not intended to apply to specific problems and situations, as no two can be alike. Reasonable care has been taken to ensure that the principles and methods described here are correct, timely, and relevant to industry today. However, no representation is made as to their accuracy, completeness, or suitability for specific situations; it is up to the reader to investigate and assess each situation or problem on its own and to apply strategies devised specifically for that set of circumstances, conditions, and requirements.

Charles E. Witherell

Mechanical Failure Avoidance

Chapter

1

Failure: Misfortune or Avoidable Event?

What Is a "Failure"?

Entropy and forestalling the inevitable

No one wants failures. They need not be major disasters. Petty annoyances such as broken shoelaces, flat tires, or split fingernails can nearly spoil our whole day. We have been conditioned to avoid failures since our infancy. Even trivial everyday failures make us discouraged and irritable. And their negative effects upon us can far outweigh their seriousness.

Before we tackle the problem of mechanical failures and practical strategies for avoiding them, we must define our targets so that we can recognize and evaluate the effectiveness of our efforts. For openers, we have to face the fact that failure *prevention,* in the absolute and practical sense, is not achievable. "Zero defects" is a noble objective but, unfortunately, an unattainable one in the world in which we live and work.

Failure *avoidance,* on the other hand, is a more reasonable and realistic goal. It admits the inevitability of failures and concentrates upon how they occur and how to keep free of them. Probably a better term for what we are dealing with here is *forestalling,* or thwarting, failures. All this is more than a game of semantics, for, unless we have our priorities in order, we can end up spinning our wheels instead of effectively avoiding failures.

But why should failure prevention and defect-free products be unattainable? The answer lies all about us. Browse through an automobile junkyard, a disposal dump, even our own attics or garages. The brutal truth is that sooner or later everything fails. We can painstakingly attend to producing a flawless design, specify the best materials and workmanship, and strengthen and preserve every component but, like Oliver Wendell Holmes's "wonderful one-hoss shay" where every part was made precisely as strong as the rest (no weak links), the product will one day disintegrate, end up in a heap, and disappear ("...it went to pieces all at once, and nothing first,/Just as bubbles do when they burst").[1] Someone has aptly remarked that Murphy was an optimist.

Engineers and physicists understand from thermodynamics that this inevitable falling apart of everything is due to the pervasive and universal phenomenon, termed *entropy,* expressed in the second law of thermodynamics. Briefly and simply stated, the law says that if things are left to themselves, disorder tends to increase. Hot coffee gets cold; beer turns flat; machines seize; and, welding and epoxy notwithstanding, Humpty-Dumpty just cannot be put back together again.

Engineering is carried out in an attempt to reverse some very limited region of disorder. We do this in many ways, but always at the expense of resources such as work, time, and energy. For example, we mine and recover ores, smelt and refine them; purify and combine natural substances; shape and assemble the refined materials into forms and products that are useful and beneficial. In doing so we have transformed disorder into order, although on a very small scale.

While we can successfully restore order to selected pockets of the disorder around us, the restoration is only temporary. In restoring order we have not revoked or superseded the second law, merely postponed its apparent effects for a while. We know that refined metals left to themselves revert back to their original ores; rot, various organisms, and oxidation of one form or another return organic materials to their constituents, and so on. These reversions to disorder can be hastened or slowed by temperature, environmental conditions, and combinations of conditions, and we can forestall such reversions to some extent through compositional modification and protective treatments.

For example, we alloy iron with elements such as carbon to enhance mechanical properties and usefulness at ambient temperatures. To resist softening at elevated temperatures, we add chromium, molybdenum, nickel, and other elements. For prolonged exposure to elevated temperatures, composition must be further controlled to minimize certain trace impurities to resist weakening by creep and precipitation of embrittling species. For more demanding service, as in turbine blades for aircraft jet engines, other, more resistant metals must be used and alloyed and, frequently, protective diffusion coatings applied to resist surface attack and deterioration.

We have developed ingenious ways to resist the ravages of exposure to hostile environments, hard use, and even misuse. But all we have really done is to buy time and prolong life somewhat. If any material or product is stressed enough, long enough, and exposed to a severe enough environment, even the "best" will fail. Everything has limits. All it takes is for the weakest link in a process, manufacturing plant, or mechanism to cease to function and the entire system or machine can grind to a halt. Nothing is "forever" despite what DeBeers says about diamonds. Although we may devise ways to delay failures, in the end everything disintegrates into disordered oblivion.

In delaying failure, engineering requires difficult and delicate balancing and inevitable compromising among the frequently bewildering array of needs, purposes, intentions, expectations, costs, and inherent risks of the project or undertaking. For example, if we make the product, structure, or system too strong or more durable than it need be, it will be too expensive, too heavy, or too cumbersome to use efficiently. Make it too slender or lightweight, cater too much to esthetics, or cut costs at the expense of other considerations, and it is likely to fall apart prematurely.

Engineering design, as we shall discuss later in some detail, is often a frustrating task of attempting to satisfy conflicting objectives, desires, and requirements. Failure avoidance requires understanding of pitfalls and penalties along the way and utmost skill in maneuvering around them while still fulfilling all the objectives.

Awareness: Prerequisite for avoidance

A principal task of engineers is to forestall failure: that is, postpone it until the device or system in question has fulfilled its intended purpose or is no longer needed. This is not always easy to do, as the intended purpose is seldom clear or definite and the required lifespan often unpredictable. However, if we start off with an intelligent understanding of what and where the traps are, we have a fighting chance of making it safely through the maze.

Someone may question why we are starting things off on a negative note, stressing the impossibility of preventing ultimate failure. It is because we must set our sights upon realistic and achievable goals. As engineering tasks become increasingly complex and there is intensified competition for available resources, we cannot afford to devote our energies to the pursuit of unattainable goals. If we are to avoid costly and damaging failures successfully, it is essential that we understand how and why they occur so we can apply reliable techniques to avoid them.

Effective strategies must start with knowledge of what it is that must be avoided. This holds true in avoiding sickness, disease, crime, and poverty, as well as engineering failures. Public service messages on how to keep from becoming a target of crime or a victim of the HIV virus or on other dangers stress the importance of maintaining an enlightened *awareness* of the risks out there and show how certain kinds of behavior can increase these risks and lead to serious trouble and even disaster. Similarly, the message here is to understand the forces that are relentlessly at work to make your engineered product fail before its time. Armed with this knowledge, it is possible to apply principles of failure avoidance and achieve the goal we all are seeking—engineered products, processes, and systems that reliably fulfill their intended purpose.

Defining "failure"

Most of us perceive failures, especially mechanical failures, as spectacular events wherein something suddenly breaks, ruptures, or falls to pieces. These are, of course, failures, but they do not necessarily represent the majority of failures or the most significant. Usually, such dramatic incidents are but the culmination of a series of subtle and insidious events that began some time before. Were it not for these earlier, largely ignored, events the final spectacle might not have taken place. In our discussions we will be stressing the critical importance of these "preliminaries": the conditions, decisions, situations, behavior, and attitudes that so surely set the stage for the "main event."

For engineering to be carried out intelligently and effectively, there must be some understanding of the purpose and intent for the "ordering" that we propose to do, some good reason for expending the effort, energy, and other resources. This is the reason there are standards, goals, specifications, and targets. These must take into account the purpose and intent of the engineering undertaking. Recent developments in products-liability law have given special emphasis to the *expectations* of those who will ultimately come into direct contact with what we will do, make, or say or be indirectly affected by it. Failure, then, is any missing of the mark or falling short of achieving these goals, meeting standards, satisfying specifications, fulfilling expectations, and hitting the target.

Although the word *failure* in engineering context usually brings to mind spectacular and devastating occurrences that make headlines, these are relatively rare events and account for a very small percentage of the failures that concern us. The dictionary definition of *failure* is relevant here, as its primary meaning is to fall short in something expected, attempted, or desired, or to be in some way deficient or lacking.[2]

Failure today, in the context of what engineers do, must be defined broadly. It can be any incident or condition that causes an industrial plant, manufactured product, process, material, or service to degrade or become unsuitable or unable to perform its intended function or purpose safely, reliably, and cost-effectively. The definition should even include operations, behavior, or product applications that lead to dissatisfaction, unattained goals, or undesirable, unexpected side effects.

This broadened definition has been imposed upon engineers with all the authority and force of a new engineering standard. It has originated within our society and reflects its increasing concern for the welfare of its members and its growing intolerance for harm and injuries caused or perceived to be caused by, among other things, technology run amok.

Consequences of Engineering Failures

Society's perceptions

Engineers are those entrusted with conversion of raw material and other resources into useful forms for the benefit of all. Implicit within this trust is the expectation that this conversion will be carried out responsibly: this means without hurting people physically or economically, or damaging the environment. This has been a recurring theme of engineers' codes of ethics for well over a century.

Nuts-and-bolts tasks of conversion of resources into useful products and systems to enhance our living standard lie within the engineers' province. They are, of course, not the only ones involved. Politicians, economists, lawyers and the courts, and society itself all share some responsibility and play a role in getting these things done. However, since engineers are trained in the necessary technical skills and capable of determining what is right and what is not in these matters, they are the ones who are logically in the spotlight when things go wrong.

Nevertheless, it is not unusual in the aftermath of a destructive industrial accident or structural failure for the media to present authoritative-sounding discourses by the man-on-the-street with critiques on causes and remedies. With benefit of 20/20 hindsight and strewn wreckage, uninformed laypersons suddenly become experts in stress analysis, aerodynamics, fracture and fatigue, or other equally sophisticated technologies. We must sympathize with the engineer and corporate executive who are called on the carpet before a nationwide TV audience in the aftermath of a failure. Before the dust has settled or an inquiry has gotten under way, they are expected to explain exactly what happened and how they will guarantee that it will not happen again.

The public may have it all wrong, but their perceptions and opinions—uninformed as they may be—play a major role in the development of legal standards and regulations that have been, are being, and will be applied to us as engineers. The cause or causes of the failure may not have had anything to do with design or manufacture, but these are among the first things to come under scrutiny. Whether we like it or not, and whether it is true or not, where technology is involved, the public views engineers as guarantors of its well-being. In the public's eyes, in the event of a failure, the engineer and manufacturer hold the smoking gun.

But this is not a recent development. One's obligation to answer for harm to another is found in documents of antiquity.[3,4] Civilized society has always demanded a remedy or retribution for offenses against it. This is deeply embedded within the fabric of our social values and is basic to our products-liability law of today. Whatever our view of products-liability law and our desire to see it modified or abolished altogether, indications are that it will not be going away anytime soon. Society demands it. We may dispute the reasonableness of public policy or the correctness of society's values, but we are powerless to change them.

Over a century ago Abraham Lincoln said, "With public sentiment nothing can fail, without it nothing can succeed." A leading jurist, in summarizing the purpose of the law, made this comment:

> The final course of law is *the welfare of society*.... When [judges] are called upon to say how far existing rules are to be extended or restricted, they must let *the welfare of society* fix the path, its direction and its distance.[5] (emphasis added)

Society has always taken a dim view of damage caused by someone or some activity suffered by its members. But it was previously much more tolerant of industrial accidents and damage traceable to "technology" than it is today. This earlier tolerance is echoed in an 1873 New York court decision involving an exploding steam boiler. In a refusal to hold the owner liable for the ensuing damage to neighboring buildings, the court said:

> We must have factories, machinery, dams, canals, and railroads.... If I have any of these on my lands...I am not responsible for any damage they accidentally and unavoidably do my neighbor. *He receives his compensation for such damage by the general good, in which he shares.* (*Losee v. Buchanan*[6]) (emphasis added)

At the time, society believed that science and technology were the keys to utopian prosperity that everyone hoped for. Samuel Florman describes it like this: "The conventional wisdom was that technological progress brought with it real progress—good progress—for all humanity, and that the men responsible for this progress had reason to consider themselves heroes."[7]

The law of the day even erected fictitious barriers to insulate industrial enterprises against lawsuits by injured parties. A principal fiction was known as *privity of contract*.[8,9] The substance of this concept was that legal action by a party harmed by a manufactured product could not be sustained against the manufacturer without the existence of a contractual relationship between them. This meant that only dealers or distributors, dealing directly with the manufacturer (parties unlikely to be injured by the products), could bring legal action. Injured bystanders were, of course, in the same boat as they, too, were without "contractual privity" with the manufacturer. A 1903 court decision stated the situation this way:

> [T]here must be a fixed and definite limitation to the liability of manufacturers and vendors for negligence in the construction and sale of complicated machines and structures which are to be operated or used by the intelligent and the ignorant, the skillful and incompetent, the watchful and the careless, parties that cannot be known to the manufacturers or vendors... (*Huset v. J. I. Case Threshing Machine Company*[10])

As the saying goes, "That was then but this is now." As written over an entrance to the Yale Law School quadrangle, "The law is a living growth, not a changeless code." Note the sharp contrast with the Huset case, above, in a later court decision that was instrumental in the widespread adoption of the concept of *strict liability*. Note, particularly, in the following excerpt from the court's decision, the importance placed upon public policy and public interests:

> ...it should be now recognized that a manufacturer incurs an absolute liability when an article that he has placed on the market...proves to have a defect that causes injury to human beings....public policy demands that responsibility be fixed wherever it will most effectively reduce the hazards to life and health inherent in defective products.... It is evident that the manufacturer can anticipate some hazards and guard against the recurrence of others, as the public cannot. Those who suffer injury from defective products are unprepared to meet its consequences.... It is to the public interest to discourage the marketing of products having defects that are a menace to the public. If such products nevertheless find their way into the market it is to the public interest to place the responsibility for whatever injury they may cause upon the manufacturer, who, even if he is not negligent in the manufacture of the product, is responsible for its reaching the market. (*Escola v. Coca Cola Bottling Co. of Fresno*[11])

It is evident from the above excerpt that this century has witnessed a remarkably rapid evolution in the law that has revolutionized legal standards for manufactured products and those that make them. These standards are as important as any technical standard; in some situations they can be even more so. They are standards against which everything designed, processed, manufactured, and sold in this country is evaluated. These evaluations are not conducted in our corporate laboratories in environments conducive to scientific objectivity but in courts of law. The rationale for courts' standards superseding industry's standards is described in an early, frequently cited, federal court decision:

> There are, no doubt, cases where courts seem to make the general...[industry practice] the standard of proper diligence...a whole [industry] may have unduly lagged in the adoption of new and available devices [or practices]. *It never may set its own test,* however persuasive be its usages. *Courts must in the end say what is required*; ... (*The T.J. Hooper*[12]) (emphasis added)

Products-liability law, however, is but one sign of society's changing values. Virtually everyone in industry today is aware of the multiplicity of governmental and administrative regulations that cover every aspect of operation. Not only must each incoming raw material and component have accompanying safety assurance documentation, but the same and more are required for finished products, scrap, by-products, process and material wastes, effluents, spent reactants, and residues. Similar regulations exist for operations other than manufacturing, such as mining, refining, bulk storage and transport, and disposal of practically everything.

Technology: Root of all evil?

What is the reason for this flood of regulations and demands for accountability by industry? Is it the quest for safety and the protection of society, or is it a reflection of antitechnology sentiments of recent years? In the futile search for a simple answer we are reminded that—like most engineering failures—there probably is no single simple cause. We live and operate in a complex world and we cannot expect that the tide of public sentiment and sociopolitical policies toward the technology that pervades our existence can be stagnant.

Incentives for the antitechnological "movement" are examined from an engineer's viewpoint in Florman's book,[13] *Blaming Technology—The Irrational Search for Scapegoats,* and it offers interesting insights into the changing tide. Florman, in admitting its complexity, suggests that the recent wave of criticism directed toward technology may be founded upon a fear of it. He observes that:

> The current mood of apprehension [over "runaway" technology] has percolated through our social consciousness in a diffuse pattern that is almost impossible to trace. It starts with disasters, but...there have always been disasters. It is the way we respond to disasters that has changed.[13]

During the several decades that have seen the changes we have described, others also have taken place. It is instructive to consider briefly some of those most relevant to engineering failures and their consequences.

Standing high, probably near the top of the list, is technological dependence. It is difficult to name some phase of our daily lives that does not rely heavily upon some technological product, service, or activity. Even homeless derelicts own transistor radios. "Manual" labor has virtually disappeared. Laborious tasks are almost exclusively done with motorized or engine-driven implements—from leaf (and dust) blowers to ditchdiggers, shovelers, rakers, mowers, weedwhackers, saws, snowblowers, and pneumatic nail drivers. Cordless appliances are found in every niche of our households. Computers have become indispensable to our daily affairs from supermarket checkouts to banking transactions, telephone communications, writing, recordkeeping, automotive systems, and video games. From the food we eat, the clothing we wear, to how we spend our leisure moments and maintain our health, everything depends upon technology.

This could have been said several decades ago, however. Yet, the sophistication of today's technology distinguishes it from that of the 1940s, 1950s, or even 1960s. For example, few people—even those of us working in scientific fields—can explain the construction and operation of the common devices we use every day. Try to repair a balky videocassette recorder, pocket stereo receiver, cellular telephone, or the ignition or fuel distribution system of your late-model car. Materials of construction and methods of assembly and disassembly have become obscure.

The printed admonition on many products that there are "no user-serviceable parts inside" the device means to forget about trying to fix it yourself. Here are your options: discard it in the trash and buy a new one, or find someone skilled in its peculiar technology with access to service manuals, special tools, and spare parts to repair it—that is, if you are willing to pay a sizable fraction of its replacement cost merely to have your old unit working again (and for how long?). These experiences intimidate purchasers into reluctantly considering the added costs of extended warranties the next time they are faced with laying out a bundle for some technologically sophisticated device. While discarding faulty products and replacing with new is often the only reasonable alternative, these limited options do not create a favorable impression in the minds of the user-owner. It always seems as though the thing should have lasted longer, and the nagging suspicion that its obsolescence was cold-bloodedly planned does not help.

In recent years we have become conditioned to accept the disposable product (of course, everything is inevitably disposable). Batteries, flashlights, pens, lighters, even cameras, not to mention fast-food and drink containers, diapers, razors, and medical implements can be disposed of after one use. Many of these are giving headaches to environmentalists and disposal site operators.

Introduction of new materials and the growing list of them in every industry are a source of problems in themselves. Their storage, transportation, and disposal lead to concerns over accidental releases and contamination. Major highways and waterways are frequently closed, awaiting identification and recommended clean-up procedures for some mystery substance that has leaked from its container in transit. The news media report evacuations of people from buildings, schools, workplaces, hospitals, and residential areas in fear that fumes, odor, or contact with some known or unidentified substance that has been inadvertently released may be harmful.

The need to maintain and transport sizable inventories is the inevitable result of widespread use of new processes, materials, and their reagents and constituents. If the materials are toxic or hazardous in some way, the risks of damage and injury increase significantly. Recent accidental releases associated with chemical plants and oil tankers, and the public's outrage over them, are indicative of the potential problems inherent in large-scale operations.

Accompanying any new technology, new materials, new assembly and construction methods, and their increased adoption are the increased risks of failure and resulting damage. Is this new technology thoroughly understood? What are the consequences of the inadvertent release of large quantities of some new chemical upon the population, wildlife, and environment? How about long-term effects and side effects? The media seem to delight in reporting accounts, and even horror stories, of such incidents.

Each such event, and many seemingly trivial incidents, must be considered failures according to our broad definition. The occurrence in question may not have caused apparent harm, no one may have been injured, no property may have been damaged, and costs may have been negligible. But, if the incident was *perceived by the public* as a failure, then it *was* one and must be treated so by responsible engineers, managers, and technical personnel. Because of the growing intolerance for engineering failures, the expanding scope of what constitutes a failure, and the litigious nature of our society, engineers today—as never before—must continually assess and reassess the potential effects of everything they make, do, or say.

"Sticker shock"

The need to accommodate these changes in society's values, expectations, and sensitivities is costing us all enormous sums of money. Its realization can be most acute to those still operating under yesterday's standards and values, unaware of or blissfully ignorant of the new ones that society and the courts have imposed upon them—that is, until its costs are pounded home by failure to comply with a regulation or the need to defend a lawsuit.

The bill can come in many forms: tightened governmental regulations and increased bureaucratic interference, broadened and intensified environmental and safety legislation, stiff fines and economic sanctions for noncompliance, litigation costs and settlements, higher insurance premiums, product recalls, and eroded goodwill and investor confidence over adverse publicity or diminished earnings. This is topped off by escalating taxes to pay for it all, not to mention the increased expenses that must be passed on to the customer in one form or another. The cascading cumulative effects of this at all levels have blighted corporate earnings and our nation's economy. Most of us probably have not yet fully understood or realized the total impact.

Have all of these developments taken place because of engineering deficiencies? Probably not. Failures occur for any number of reasons, as we shall see in the next section. Many have nothing at all to do with integrity of the engineering design or manufacturing or material flaws. But if an engineered structure or product is involved, it is often guilt by association. It is human nature to fear the unknown. Because the average layperson is mystified by what engineers do and make, he or she tends to draw inferences and sweeping conclusions based upon ignorance, for the most part, and fragmentary knowledge, at best. As long as the public tends to view engineers and technology as a "loose cannon," it is probable that failures of technical things will continue to be placed at the feet of those believed to be responsible.

Much of the technologically related environmental hysteria in recent years—objections to nuclear power, for example—is founded on incomplete and often incorrect facts. Emotionalism generated over incidents perceived as proof that the system is inherently flawed, with impending disaster just ahead, is largely the result of agitation by activists having their own agendas. Unfortunately, when it comes to technical matters, the public cannot sort out fact from fiction and tends to buy into their arguments.

Fear of the unknown leads to skepticism over technology. Soon, issues hostile to technological progress take root in the form of tightened regulations and adverse court decisions. It's an ever-tightening spiral. We cannot change the past, only try to be smart enough to avoid reliving it. We have seen what a generation or so of suspicion of technology has done to our industrial and social climate. Some of it is for the best, of course. But there has been too much skepticism, unfounded hysteria, and emotional overreaction. Unless we do all we can to counteract this type of negativism, it will become much worse.

How can we do this? We can start by minimizing failures. It cannot be done haphazardly or sporadically, but responsibly and consistently.

It may mean looking at what we do in a different light, adjusting priorities and values, placing more emphasis upon educating the public, becoming more directly involved in hands-on solutions of society's problems, and acting responsibly 24 hours a day. We must assure society that what we do and, more significantly, what we are perceived by it as doing can withstand the tightest scrutiny.

The Causal Web and Interrelated Complexities

Causation categories

Failures occur for a host of reasons. Even simple and relatively inconsequential accidents and failures invariably have an intricate network of causation, that is, the circumstances that led up to them. The network is comprised of errors (of commission as well as omission), wrong attitudes and incorrect beliefs, misplaced confidence, scrambled priorities, defective communications and information saturation or overload, ignorance and uncertainty, conceptual misunderstandings, and confusion of roles and assignments. These are so-called people-related elements.

Then there are external causes, those related to surroundings or environment: material incompatibilities, unforeseen conditions, unexpected deterioration, unfavorable response, inadequate control over operating conditions, excessive demands, and unrealistic expectations. The causation network also includes faults, defects, glitches, and flaws that are found in virtually everything—some more serious than others. Within this category of causes are malfunctions, improper or inadequate maintenance, design deficiencies, and use of wrong or defective materials or components. The list goes on.

Certainly, all causative elements would not be found in every failure, even in major catastrophes. But it is not uncommon to find some of each category (that is, evidence of people-related, environmentally influenced causes, and inherently flawed objects or systems) in most failures.

We should admit here at the outset that these categories are somewhat contrived, as essentially all of these elements, in all categories, in some way involve or stem from human deficiencies, negligence, and other people-related shortcomings. For example, environmental conditions may be identified as the primary cause of a given failure, but perhaps these conditions should have been foreseen by someone in charge, perhaps the structure or object should not have been trusted under such conditions, or it might have been better designed to withstand them.

All of these are people-related deficiencies. Likewise, people-related human faults in one form or another may be found to lie at the root of most failure causation elements. Nevertheless, it is helpful in discussions of failure causation, and in the development of avoidance strategies, to attempt to categorize the elements according to their dominant characteristics.

The causation scenario has been depicted in a number of ways: as spokes of a wheel, legs of a stool, concentric layers (structured like an onion), as a web, and by logic networks. The idea in all of these depiction attempts is to convey the interrelated nature of causation elements, and that there can be many of them having differing origins but each contributing materially and converging upon the final event, or failure. Often, elimination of even one element can stifle, or at least delay, the failure. And there is the crucial role of timing, which is difficult to depict because the elements must converge at some point in time to precipitate the failure event.

Developing causal awareness

The particular format used for representing causation is probably unimportant. What matters is that we have some means for helping us picture in our mind's eye the interrelated nature of the causative elements, of all kinds, and for identifying and cataloging them in an orderly and complete manner. This is needed to assist us in learning from previous failures, the mistakes of others, and future blunders committed by others as well as ourselves. There is tremendous value in learning from others' failures—whether they are in our actual field of specialization or not.

There are a number of ways one might approach failure reports. If we regard them superficially as merely news items of some remote disaster or misfortune, or as a passing interest, as we might the massacre of coworkers by a disgruntled and deranged gunman or tidal wave devastation of coastal villages along the Indian Ocean, we will have missed the value and opportunity for learning from the mistakes and failures of others. But if we analyze available details as they emerge and begin to recognize common causative patterns and links, we will become better equipped ourselves to avoid failures within our own spheres of engineering activity and responsibility.

What should we be looking for? The normal reaction is to concentrate upon the final dramatic event and its aftermath, but much more can be gleaned from examining the "preliminaries": the series of events, occurrences, conditions, and situations that led up to the critical moment of failure. Try to identify those ingredients that, had they been absent from the scenario, would have prevented the occurrence from happening as and when it did. Similarly, identify elements essential to the occurrence.

Exercises such as these can develop skills and create levels of awareness that can pay dividends in our own engineering world. Also, try to identify, as specifically as possible, features, elements, or characteristics common to several different failures, particularly for those within your field of expertise or company's operation. Then, it is useful to identify examples of all types of failures within our broadened definition—including failures that may not have been catastrophic in a physical sense (involving fractures and the like) and look for common threads running through them.

When it comes to failures of a type that might have occurred in *your* company, with *your* product, within *your* sphere of responsibility, a much more intensive analysis needs to be done. One of the first questions to ask is if it could have happened to you, your company, or your product. If so, what is preventing it from occurring? What differed from the situation where the failure occurred and that of your plant? Conduct the same exercise as with other failures, except with deeper scrutiny. Find out more details and keep the inquiry going until you are satisfied that you have identified causative elements that do not exist in your situation or until you institute steps to assure such occurrences will not happen with your product or plant.

These kinds of analyses can be conducted any time; we need not wait around until one is reported somewhere. There are volumes of published case studies and proceedings of symposia on the subject that deal with failures of all kinds.[14–19] Most of them will involve losses having serious economic consequences; not all contain sufficient detail on events and conditions leading up to the failure, and some may have incorrect diagnoses and solutions. However, usually there are enough details and illustrations about the incident to be worth studying. And more often than not, the conclusions are technically sound.

A word of caution concerning the use of case studies and reviews of someone else's accidents and failures: Since no two failures (or causation networks) are alike, it can be misleading at best to attempt to adapt the diagnosis and solution to other *seemingly identical* situations. It is always best to stifle this temptation and conduct your own study of your own materials, processes, and designs, and draw your own conclusions. Case histories have value in the context of our discussion on learning to identify causation elements; they are not "instant cures" for what ails your product or operation, and they are not shortcuts to a properly conducted in-house investigation (see Chap. 2).

Besides case histories, another fruitful source of failure information can be legal transcripts and written decisions of civil litigation appeals arising from engineering and other failures. Since most lawsuits are settled before going to trial, a good many of them will not be found. Also, courts are not necessarily interested in analyzing the entire spectrum of elements of failure causation—only those that impinge directly upon the legal issues before them. Nevertheless, many of these written accounts (found in law libraries) contain details of prior failures that were sufficiently significant to warrant a lawsuit and appeal and, therefore, are valuable resources for study.

Intrinsic Susceptibilities and Extrinsic Inevitabilities

Some of the more relevant aspects of failure causation covered in the foregoing discussions are illustrated in the following three failure examples, selected almost at random. They involve a leaking gasoline tank aboard a pleasure boat, a helicopter searchlight that tore loose from its mounts, and a laboratory experiment that went awry. None of them involved litigation or serious injury, although the potential clearly was present in all three.

These are not presented to demonstrate proper investigative procedure, clever diagnosis, or even failure avoidance techniques. They are presented merely to show the complexity and frequent implausibility of the web of causation surrounding common failures and how well intentioned but sometimes incorrect decisions and assumptions can so easily create a failure scenario that is virtually inevitable.

The perforated gasoline tank

The first example involves a welded rectangular-shaped 150-gallon gasoline tank $^1/_8$ in (3.2 mm) thick of type 5052 aluminum alloy, fabricated of flat sheet with brake-formed radiused corners and flat slanted bottom (to conform to the hull contour). It had been installed for a few years in a pleasure boat operating out of warm coastal seawater ports.

The odor of raw gasoline led the boat operator to search for its source. After fittings and lines were found secure, the tank was removed and a sizable irregularly shaped perforation completely through the wall of the tank was observed at the lower corner where the welded seams joining the tank end to the formed bottom corner intersected. See Fig. 1.1. The edges of the perforation had a granular texture and a dull-gray color, suggesting localized corrosive attack. The region appeared to have been previously weld-repaired. The rest of the outer surface of the tank appeared sound, intact, shiny, and unaffected.

When the tank was cut apart to inspect the interior, a light deposit of brown-colored residue was observed on the inner wall, in and for a distance of several inches around the perforated corner. The fuel draw tube, later determined to be of copper, was located near the affected

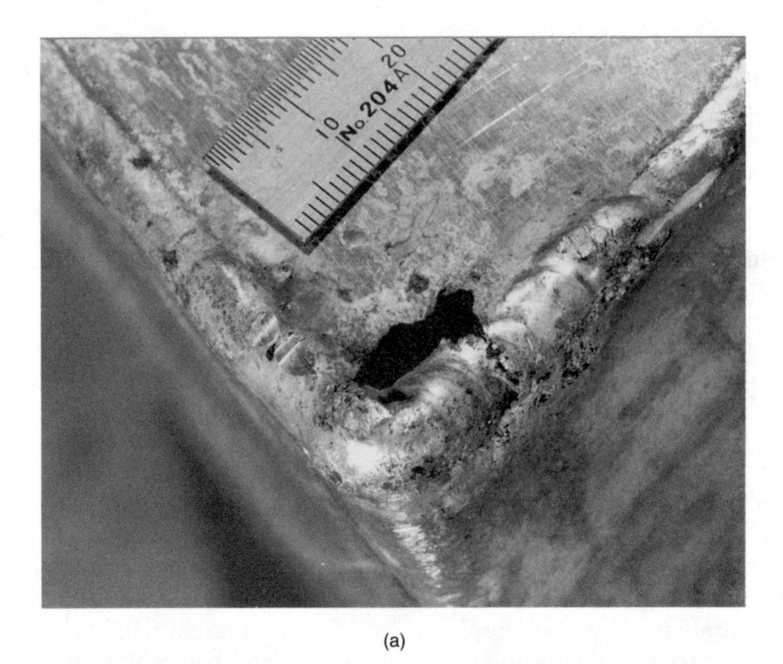

(a)

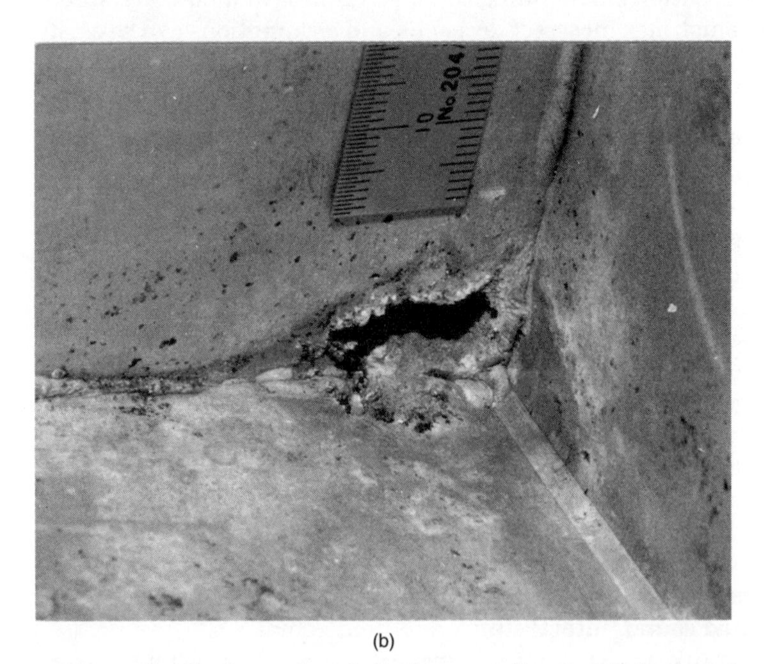

(b)

Figure 1.1 Localized corrosion of aluminum gasoline tank [(*a*) external view and (*b*) internal view].

corner, and it had a dull surface texture and coloration similar to that of the deposit on the inner wall near the perforation.

Adjacent to the weld repair, at the external surface of the tank, were what appeared to be melted and resolidified, hard, brittle, adherent, thin surface deposits of a cloudy grayish light blue-green color. These had the appearance of weld flux residues but were not, since the repair welds (as all other tank welds) had been made by the inert gas–shielded tungsten-arc process.

Interior examinations disclosed a number of other sites of localized corrosive attack on the bottom of the tank, almost directly under the draw tube (its orifice was situated about 1.5 in, or 38 mm, above the tank bottom). Attack in these other regions was relatively superficial, however. There were also brownish-colored deposits along the formed lower corner with scattered shallow pits.

Several facts were uncovered during questioning of the parties involved. The perforated corner and adjacent formed corner where the bottom and side intersected were the lowest points of the tank when it was installed in the boat. It was reportedly not uncommon for small volumes of water to occasionally accumulate there, either through the filler spout during refueling or by condensate from the humid operating environment. The use of copper for the draw tube, instead of a nonmetallic material that was commonly used, was reported to be a premium option that had been selected.

Perforation of the tank was determined to have been the result of galvanic corrosion stemming from use of a copper draw tube in the presence of accumulated residual water that had collected along the V-shaped crevice inside the tank at the bottom. Under these conditions, the aluminum tank—as an anode in an electrochemical corrosion couple—would preferentially corrode. The copper tube (serving as cathode) and water (serving as the electrolyte) completed the cell.

The welded corner was a favorable site because it is a region of high residual stress due to intersection of the seam welds there, plus additional stress from the weight of the fuel above it. Weld metal and heat-affected zones would have coarser-grained microstructures than the tank wall, and the chemical composition of the weld metal itself would differ somewhat from that of the tank wall since weld filler metals for this alloy do not match the base metal composition exactly. All of these factors would tend to make the welded corner region of the tank more susceptible to corrosive attack than other regions, given the proper conditions for corrosion to occur.

Here we have a welded aluminum fuel tank that, under the service conditions involved, should not have experienced corrosion. The choice of copper for the draw tube would similarly appear to be proper, considering the well-established excellent response of copper and its alloys to marine environments. Yet, the dissimilar-metal combination in close proximity and in the presence of accumulated water at the low point of the tank provided the right set of conditions for localized attack at a particularly susceptible location. Indications are that this was not the first time the tank corroded through and leaked at this location, as there had been a weld repair there and traces of melted and resolidified copper- and aluminum-containing corrosion product (the faintly green-blue cloudy-gray residue).

What was the cause of this failure? Was it corrosion? The materials used had a good service record in such environments. The copper tube was chosen over one of plastic to give added assurance of years of trouble-free service. Was the decision to use copper for the draw tube the cause? Even if the galvanic phenomenon between copper and aluminum had been known by the parties involved, was it reasonable to expect that galvanic corrosion would be a problem—within a tank of gasoline?

Should they have foreseen the possibility that water, accumulating in that region and becoming an electrolyte for galvanic corrosion between the copper tube and aluminum tank, might lead to perforation of the tank wall and a potentially dangerous gasoline leak? Was the water the cause? How about that earlier leak or leaks and the attempt to weld repair them, which was apparently successful for a time? Should the welder or repair facility have recognized that localized corrosion was really the culprit and that the weld repair might not really solve the problem?

Note, in this relatively simple example, the fairly complex web of causation. People made choices and decisions that were thought to be conservative and, above all, correct. No one tried to cheat, take shortcuts, or compromise. The weld repairs appeared to have been done competently. From all that could be determined, no applicable code provision had been violated. No safety practices were ignored. There apparently had not been any misuse or abuse, or untrained or careless boat operators. Yet, in spite of all of these positive factors, decisions, actions, and circumstances converged unwittingly to set the stage for what could have been a disaster.

This example is reminiscent of a debacle on a larger scale where every attempted weld in a massive copper-sheathed piece of cryogenic apparatus failed. A $^{1}/_{16}$ in (1.6 mm) thick copper sheath fully encasing a large complex-shaped unit was to be formed and welded in a fixed position to provide a leak-tight enveloping seal throughout. The welding engineer had specified a grade of deoxidized copper that would be weldable under the field conditions involved. However, someone along the way countermanded the specified grade of copper and substituted a purer grade having higher thermal conductivity and commanding a substantially higher price since it was a premium material.

As it turned out, the premium grade of copper, because of its lack of residual deoxidizers, was prone to severe weld embrittlement, so much so that an uncracked weld could not be made—with seams frustratingly separating almost as they were made. Repair attempts were futile, as all they accomplished was to add more heat, produce more warpage and stress, and cause more oxygen pickup from the surrounding atmosphere, which led to even more embrittlement. Precious time was spent fruitlessly attempting to correct the problem through adjusting welding parameters, preheating and post-heating, using supplemental gas shielding, trying other welding processes and filler metals, and modifying the welding sequence.

It took months after the "error" was realized for the premium high-conductivity grade of copper to be stripped away from the vessel and scrapped and the correct weldable grade to be ordered, delivered, fabricated into shape and fitted; only then could the project get back on track. Once again, good intentions by people "doing their best" to make the project a success, yet being misinformed in some critical aspect, had the very opposite effect.

An errant searchlight

Helicopters used in law enforcement, security, and surveillance activity are often equipped with large powerful searchlights for nighttime illumination. These are generally mounted under and forward of the nose and are equipped with remote, pilot-actuated, motorized pitch and yaw controls. These allow the searchlight to be directed at ground targets, as desired, while the helicopter is airborne.

The large cylindrical searchlight body itself is mounted to a yoke, an inverted U-shaped fork. This is suspended from the helicopter fuselage by tubular members. In yaw, the searchlight pivots about the apex of the inverted U; in pitch, it pivots about horizontally oriented axes at either side of the open end of the yoke. The pitch pivots are at the yoke itself, although the searchlight is mounted to $^1/_8$ in (3.2 mm) thick plates, offset a few inches back from the pivot points. See Fig. 1.2. Pitch motion is transmitted from a motor drive mounted on one side of the yoke through a drive shaft to one of these plates; the other side is not driven but merely pivots freely. Rotation of the plates about their yoke pivot points provides pitch motion to the searchlight mounted to the plates.

The searchlight body of this example was mounted to the $^1/_8$ in (3.2 mm) thick attachment plates by two round countersink flathead 8 mm (0.3 in) steel bolts (Phillips-type recessed drive), one on each side. Four

Figure 1.2 Lower portion of helicopter searchlight yoke showing mounting plate at motor drive side.

smaller (4 mm, or 0.16 in) screws were situated around each of the main attachment bolts to hold the mounting plates to the searchlight body and maintain the proper alignment of the attachment plate. The weight of the searchlight, however, was entirely supported by the two 8 mm (0.3 in) steel bolts.

The incident report stated that the searchlight of this example had been torn from its mounts during a hard landing or collision with the ground at night. During the incident, the searchlight became completely detached from its yoke and was driven rearward into the nose of the helicopter, which sustained moderate damage as a result. Two of the four yoke attachment support tubes had broken and separated from the helicopter. The searchlight assembly was severely damaged and considered a total loss.

Examinations of the damaged components and fragments retrieved from the accident site disclosed some unexpected facts. From gouges and scrape marks on the searchlight body, it was apparent that it was torn from its mounts by a force vector in a *forward* and *downward* direction, not backward and upward as it would have been if the searchlight had collided with the ground during an uncontrolled or careless landing. This finding indicated that the searchlight had been wrested from its mounts by reaction forces (or searchlight inertia) resulting from a force vector applied in an upward and backward direction *to the helicopter*. This would occur if the helicopter contacted the ground while in a forward descent.

These findings showed that the searchlight had remained secure in its mount up to the moment of impact and became torn from its mount when the helicopter, not the searchlight, impacted the ground. Once torn loose, the searchlight was free to strike the nose of the helicopter and become ensnared with its electrical cables, causing the observed damage to the yoke and other components.

At this time, and with no additional evidence, it would have been reasonable to conclude that the damage was the direct result of a hard landing. But there was other evidence. Someone at the scene of the incident had retrieved two boltheads along with other debris and fragments. The boltheads were matched to the fractured mounting bolt stubs still in place within the searchlight body and were, in fact, identified as the heads from the two 8 mm (0.3 in) mounting bolts.

Examinations of the boltheads disclosed that the countersink-type heads had been undercut, in a machining operation, to permit them to match the countersink holes in the $^1/_8$ in (3.2 mm) thick attachment plates and do so without their extending above the plate surface. That is, the machining removed the lower portion of the countersink taper, from the unthreaded shank portion upward toward the bolthead, such that the remaining bolthead thickness at the countersink matched the $^1/_8$ in (3.2 mm) plate thickness. Note Fig. 1.3.

The purpose of this modification appeared to be to provide clearance because of tight tolerances between the searchlight body and the yoke forks. Any extension of the boltheads above the mounting plate surface would have interfered with pitch motion. While undercutting the bolthead permitted unimpeded pitch motion for the searchlight and a good fitup in the mounting plates, it also severely reduced the load-carrying capacity of the bolt, as shown in Fig. 1.3.

It was apparent that the bolts had not been manufactured in this undercut configuration, as the machined regions were fresh unplated steel surfaces whereas other areas of the boltheads appeared to have been cadmium plated and chromate-passivation treated to resist corrosion and other forms of deterioration from atmospheric exposure. These are common practices for aircraft applications.

High-resolution examinations of the bolthead fractures (i.e., the four small segments remaining within the boltheads after the machining operation) using scanning electron microscopy (SEM) tech-

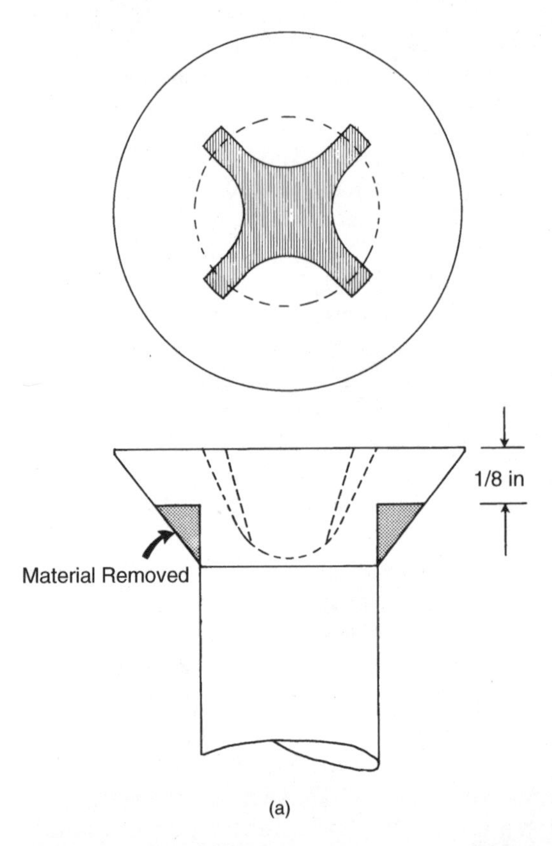

(a)

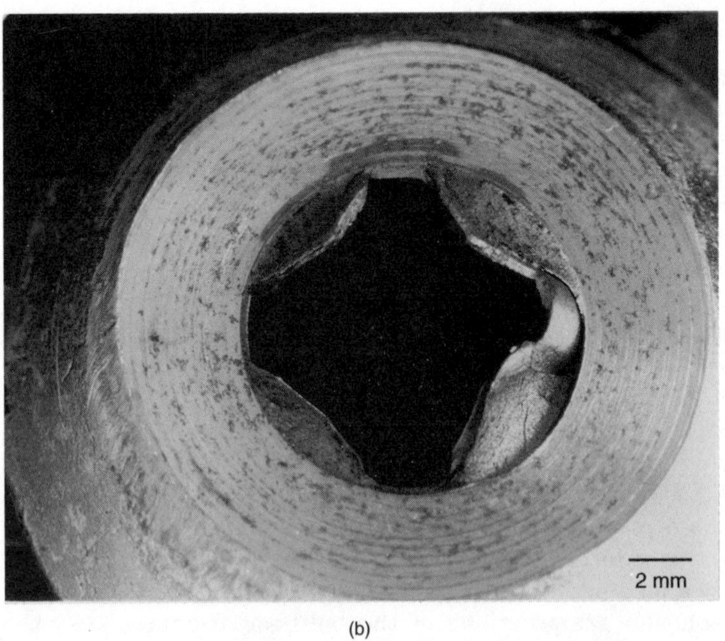

(b)

Figure 1.3 Bolthead modification to accommodate searchlight mounting plate thickness (lower photograph shows fractured bolthead as viewed from underside at surface of machined undercut).

niques showed the fractures to have a brittle appearance. See Fig. 1.4. This could have been the result of inferior chemical composition, improper melting practice during the manufacture of the bolt steel, incorrect heat treatment, improper plating procedure, or failure to perform a post-plating bakeout to remove inoculated hydrogen from the plating operation. More likely, perhaps, the bolts may not have been of a proper grade, as they had no visible markings or grade identifications.

In summary, then, machining modifications made to the undersides of the heads of the searchlight attachment bolts to permit their use in this particular assembly seriously thinned the cross section of the region under the bolthead, thereby reducing its load-carrying capability. Besides this, it produced a sharp-edged, stress-concentrating notch in the most unfortunate location of the bolt where tensile and fatigue loads are greatest. Also, the machining exposed unplated and, therefore, unprotected steel to atmospheric deterioration.

Retrieval of the fractured boltheads from the accident site played a major role in determining what had really taken place during this incident. From the appearance of the fractured bolt fragments, it is clear that very little force would have been required to fracture the thin regions under the countersink heads that were supporting the full weight of the searchlight body. A hard landing or collision with the ground certainly would not be required. Normal operating vibration may be all

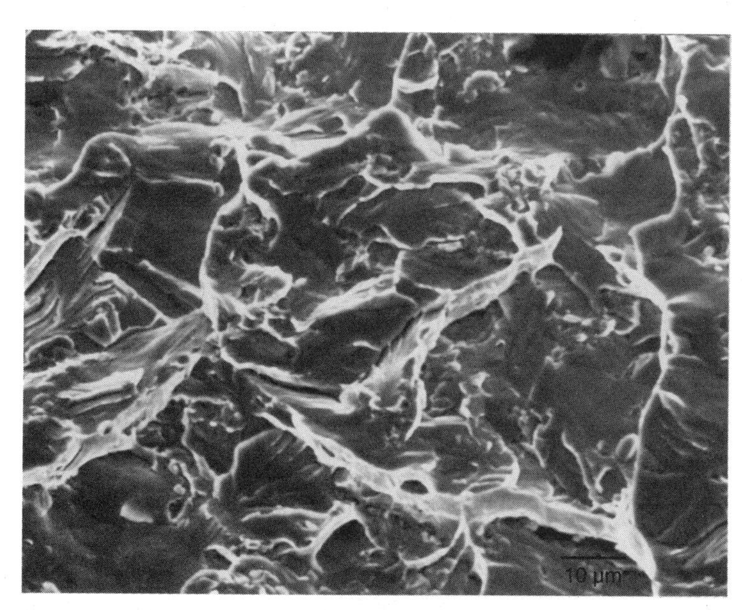

Figure 1.4 Brittle appearance of fractures in searchlight attachment bolts (scanning electron photomicrograph).

that was required to fracture these weakened bolts; however, the reports indicated that the incident occurred during landing.

The machining modification to the attachment bolts must be considered a major contributor to this incident. Nevertheless, there is the nagging question of why this machining operation was necessary in the first place. Did not the searchlight manufacturer design and supply appropriate attachment bolts? What happened to them? Why use makeshift bolts when proper ones might have been obtained? Did not the machinist consider that metal removed in that region of the bolt would significantly weaken it?

Then what about the embrittlement observed in the bolt fragments? Was this the result of exposure of unprotected surfaces to the flight environment? Was it a reflection of some deficiency introduced during their manufacture? Or were the bolts unsuitable for aircraft use altogether? If the bolts had been of the proper quality would they have survived, despite the machining modifications? How many more installations like this are out there in service? There would seem to be no end to the questions that might be asked as a result of this single incident.

Finally, there is the issue of blame and how the findings of an accident investigation affect it. If the boltheads had never been retrieved, the pilot might have been branded incompetent and irresponsible for causing this incident. It makes one wonder how many have lost jobs and jeopardized careers because of incorrect conclusions based upon incomplete facts sincerely believed to be complete but lacking in some critical detail.

The exploding canister

Hot isostatic pressing (HIP) is a process used for densifying powders into fully dense solids. The process has other applications, but the one discussed here is dominant. Powders to be densified are usually first pressed mechanically within closed dies to eliminate as much entrapped air as possible. Then the pressed compact is placed into a canister or can—usually a welded cylindrical container closed at one end.

After the powder compact is in place, a top, having a small-diameter evacuation tube welded into it, is welded on. The can is evacuated to a high vacuum and is frequently heated during the pumpdown to assist in evaporating residual moisture and other volatiles. The evacuation tube is pinched off and sealed with the can interior under vacuum. It is then ready for the HIP or densification process.

This is carried out within a heavy-walled gas pressure vessel that can be heated to high temperatures. The pressurization medium is usually argon, an inert gas, and HIP cycles typically involve exposures at temperatures up to about 1000°C (1832°F) and pressures up to about

1000 atm or more for several hours. The pressure, temperature, and time parameters depend upon the material to be densified and other considerations. When the HIP cycle is completed, the temperature and pressure are gradually decreased to ambient. The can with the now-densified material within it is removed, the outer can stripped from its contents, and the densified material further processed, formed, or machined into the desired shape. After the HIP process, the densified material can be handled and treated as the fully dense solid it is.

The incident of this example involved a laboratory installation using a small type 304 stainless steel can having a wall thickness of 0.035 in (0.9 mm) and a diameter of about 2 in (50 mm), about 3 in (76 mm) long. To ensure against possible contamination and erroneous results, it was decided to isolate the experimental powder compacts from contact with the stainless steel can. This was done using an "inert" graphitic sleeve around the contents to provide the desired barrier. All other can preparation steps were carried out according to the usual recommended procedures.

The experiment, as done, had been repeated numerous times over the previous few years without mishap and was considered a safe routine. A small percentage of previous cans did experience leaks, a condition observed after the HIP cycle when it was noted that the powder contents within the cans did not densify. But there had been no reportable incidents associated with previous runs.

In this particular HIP cycle, several nearly identical cans were exposed to the same cycle at the same time within the same vessel. At the completion of the HIP operation and after the vessel had returned to ambient conditions, it was opened to remove the densified (HIP processed) cans. The operator briefly glanced into the open vessel and noted that one of the cans had fallen to its side because its bottom was bulged and it did not have the typical "hourglass" shape of densified cans. An instant later, the bulged can exploded, scattering its undensified powdered contents and other debris over a wide area. Just prior to the explosion the operator had fortuitously turned away and was out of the direct path of the blast, thereby avoiding serious injury.

The incident raised many questions. Of primary concern was the origin of the internal pressure within the can and if some chemical reaction within the can produced it. Second, the fractures in the can fragments were brittle in appearance. Type 304 stainless steel is ordinarily a ductile alloy, from cryogenic to boiler superheater temperatures, and the observed brittleness was puzzling.

The study concluded that argon, the pressurizing medium, had entered the can through a minute leak, probably somewhere in the remaining stub of the nickel evacuation tube, as it showed signs of embrittlement (attempts to identify the cause of this embrittlement were inconclusive). If a leak path existed from outside the can to its interior, the can and its contents would not densify but remain at equilibrium with the vessel atmosphere during the HIP process.

All evidence indicated that the can had been completely evacuated before the HIP process. But the can did explode from internal pressurization, and traces of argon in the powdered debris from within the can were detected by energy-dispersive spectrographic (EDS) techniques during later examinations.

It is informative, in explaining how argon could enter the can through even the most minute leak, to recognize the effect of large pressure differentials upon leak flow rate.[20] Leakage flow is proportional to *the square* of external pressure. Leak tests on the evacuated can were conducted under a pressure differential of only 1 atm. Therefore, a leak below the detectability limit at 1 atm before being placed in the HIP chamber could experience leak flow rates as high as 10^6 times greater(!) during a HIP operation conducted at a pressure of 1000 atm.

The thin-walled canister was not intended to withstand internal pressures, and its bursting was not surprising under the circumstances—but why a *catastrophic* explosion? And why were the fractures so brittle? During examination of the fractured pieces, it was observed that the internal walls of the can had become severely embrittled, apparently from contact with the graphitic barrier material (note Fig. 1.5).

From metallographic study and other examinations, the embrittlement observed was identified as the phenomenon known as *sensitization*. It occurs in austenitic stainless steels (like type 304) during exposure to temperatures from about 500 to 1000°C (932 to 1832°F). The HIP temperature was within this range.

At these temperatures, carbon is far from "inert" and diffuses readily toward chromium (type 304 stainless steel contains nominally 18% chromium). The diffusion occurs particularly along grain boundaries and forms chromium carbides and complex carbonitrides. This condition, occurring in stainless steels having a carbon content greater than as little as 0.04%, is notorious for causing susceptibility (or "sensitivity") to intergranular corrosion in aqueous environments because of depleted chromium in these regions. (It is its chromium content that imparts corrosion resistance to stainless steels.)

While sensitization from low-carbon residuals within the steel is detrimental to the material in corrosive environments, it is not particularly detrimental otherwise—at least at ambient temperatures and above. Loss in ductility in austenitic stainless steels due to sensitization usually is not detectable above cryogenic temperatures.

When sensitization is extreme, however, as when there is an outside or supplemental source for carburization (as in this case), the effect

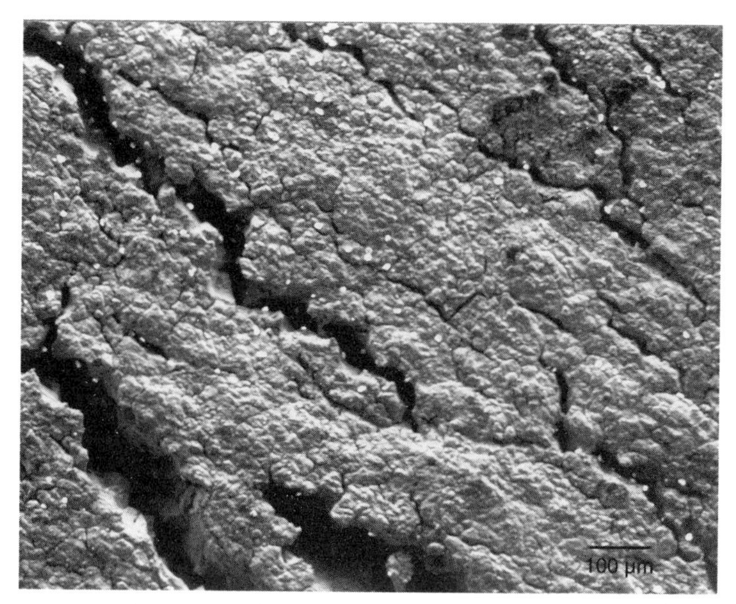

Figure 1.5 Appearance of embrittled inner surface of type 304 stainless steel HIP canister (scanning electron photomicrograph).

can be much more dramatic. The brittle and interconnected network of chromium carbide can degrade the room-temperature properties of the steel. This is particularly evident in a drastic reduction in fracture toughness and ductility or in its ability to sustain tensile deformation. Figure 1.6 shows the embrittling effect upon the fracture of the vessel wall.

In this incident, effects of two undetected conditions converged to produce the canister explosion. Without either one, it is probable that the explosion would not have occurred. As demonstrated by previous successful experiments using the graphitic liner in similar stainless steel cans, embrittled cans (although not known to be such at the time) did not cause failures. Also, leaking containers had not previously exploded but merely failed to densify. But, combine the two conditions—particularly when the leak was beyond the detectability limit of the test equipment—and the trap was set for a potentially serious accident.

Good intentions are not enough

In these three examples we have seen how potentially serious failure scenarios developed from good intentions and, ironically, often were intensified by the desire to avoid inferior or compromised outcomes. The people involved were striving for success. Yet, they failed to recognize

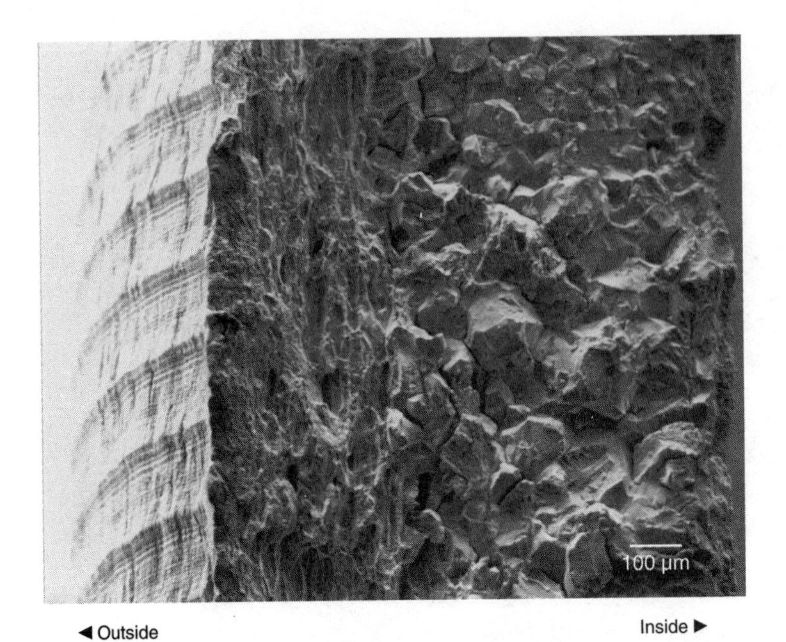

Figure 1.6 Fractured wall of type 304 stainless steel canister embrittled by sensitization at inner surface (scanning electron photomicrograph) (outside and inside directions indicated).

a technical subtlety or to understand the potentially damaging outcome of what they were doing.

Should we be so critical of them, even when it is clear that their actions and decisions set the stage for the failures that occurred? With the benefit of knowing what did occur, it is not too difficult to reconstruct the scenario and point out the deficiencies and what should have been done. It is another matter to foresee all this ahead of time and to take action that will forestall the occurrence. It has been said that hindsight is always 20/20. Would we have done differently at the time if we had been in their shoes and were operating under the same circumstances that they were when those actions were taken and decisions were made?

None of these examples might be called "major" failures. They could have had more serious consequences, but none would have made the headlines. Their outcomes were correctable with time and money. Nevertheless, their causative elements differed little from those of major accidents, failures, and catastrophic disasters; they merely involved different subject matter.

Victor Bignell distinguishes catastrophes from other types of failures simply as those that involve large amounts of energy or material on the rampage, large numbers of people unfortunate enough to be in the way, and large sums of money to set things right again.[21] He notes other common features to be magnitude, suddenness, and violence.

It would appear, then, that if we developed an understanding of causation and a sharpened awareness of how various elements can interact to precipitate a failure, we would have a solid base for applying avoidance strategies to *any* engineering situation and even nonengineering situations, for that matter.

While our goal is to avoid failures altogether, it may be too soon at this stage of our engineering development to ask or expect this much, that is, considering our limited understanding of causation scenarios and early states of development of avoidance strategies and techniques. It may be, as suggested at the opening of this chapter, that complete failure avoidance over the intended lifetime of the system is unattainable for some activities or operations—activities or operations that entail a degree of risk but are too essential to discontinue.

Air travel is one example; manufacturing and processing of indispensable but toxic materials are others. Certainly, every precaution should be taken to eliminate failures. At the same time, perhaps, we need to realize that they will occur in spite of all that is done to avoid them. Accordingly, it would be well to give serious thought, in these kinds of situations, to how injuries and the extent of damage might be diminished or mitigated during inevitable incidents.

Survivors of catastrophes frequently report that, were it not for some especially destructive or intervening element, many others would have safely made it out. Failure avoidance efforts should, therefore, address strategies for minimizing consequences of failures. It may mean only partially fulfilling the avoidance goal, but a positive step in that direction nevertheless, or until our understanding has advanced sufficiently for us to do better.

In the pages that follow, we will be considering in some detail how failures can be avoided and their consequences minimized. Techniques and strategies are necessary tools. Their intelligent use demands appreciation of the intricacies of failure causation and the key role that human beings play. It is hoped that the material presented in this opening chapter will be a catalyst in developing such an awareness.

References

1. Oliver Wendell Holmes, *The Deacon's Masterpiece,* 1858.
2. *The Random House Dictionary of the English Language,* The Unabridged Edition (1969).
3. G.R. Driver and J.C. Miles, *The Babylonian Laws,* 2 vols., Oxford University Press, New York (1952, 1955).
4. Exodus 21:1–23:9, Deuteronomy 22:1–25:19.

5. Roscoe Pound, *An Introduction to the Philosophy of Law,* Yale University Press, New Haven, CT, 1954, pp. 12, 32–34.
6. Losee v. Buchanan, 51 N.Y. 476 (1873).
7. Samuel C. Florman, *The Existential Pleasures of Engineering,* St. Martin's, New York, 1976, pp. 3–10.
8. Winterbottom v. Wright, 10 M.& W. 109, 152 Eng. Rep. 402 (1842).
9. F.H. Bohlen, *Studies in the Law of Torts,* Bobs-Merrill, Indianapolis, IN, 1926, pp. 76–80.
10. Huset v. J.I. Case Threshing Machine Company, 120 F. 865 (8th Circuit, 1903).
11. Escola v. Coca Cola Bottling Co. of Fresno, 24 Cal. 2nd 453, 150 P.2nd 436 (1944).
12. The T.J. Hooper, 60 F. 2d 737 (2nd Circuit, 1932).
13. Samuel C. Florman, *Blaming Technology—The Irrational Search for Scapegoats,* St. Martin's, New York, 1981, p. 5.
14. Shin-ichi Nishida, *Failure Analysis in Engineering Applications,* Butterworth-Heinemann, Oxford, England, 1992.
15. J. I. Dickson *et al.* (eds.), *Failure Analysis: Techniques and Applications,* Proceedings of the First International Conference on Failure Analysis, Montreal, Quebec, Canada, (8–11 July 1991) ASM International, Materials Park, OH, 1992.
16. *Metals Handbook,* 9th ed., vol. 11, *Failure Analysis and Prevention,* American Society for Metals, Metals Park, OH, 1986.
17. F.R. Hutchings and Paul M. Unterweiser (compilers), *Failure Analysis: The British Engine Technical Reports,* American Society for Metals, Metals Park, OH, 1981.
18. *Case Histories in Failure Analysis,* American Society for Metals, Metals Park, OH, 1979.
19. F.K. Nauman, *Failure Analysis—Case Histories and Methodology,* American Society for Metals, Metals Park, OH, 1983.
20. P.E. Price and S.P. Kohler, "Hot Isostatic Pressing of Metal Powders," *Metals Handbook,* 9th ed., vol. 7, *Powder Metallurgy,* American Society for Metals, Metals Park, OH, 1984 pp. 419–43.
21. V. Bignell, G. Peters, and C. Pym, *Catastrophic Failures,* 2nd printing (revised), Open University, New York, 1978.

Chapter

2

Analyzing Failures

What Past Occurrences Can and Cannot Tell Us

Learning from others' experiences

Whatever our fields and levels of experience or expertise, we continually draw upon knowledge that others have developed. This is how advancements are made. Even accomplished artists, athletes, musicians, actors, and surgeons devote much of their time studying under masters to improve and maintain proficiency. Isaac Newton wrote in a letter to the English physicist, Robert Hooke, "If I have seen further (than you and Descartes) it is by standing upon the shoulders of giants."

It not only makes good sense to build upon what others have already done; to do otherwise would be foolhardy. This is one of the principal roles of engineering societies in their sponsoring of interchanges of information at annual meetings, conferences, seminars, and in journals. Through them we find out what others have tried, have done, and, sometimes, have been unable to do. In the same vein, we engage outside specialists who have more experience or a clearer perspective on some aspect than we do.

At the outset of new projects, we conduct thorough literature searches to find out if what we contemplate doing may have been already tried by someone else. If so, we want to know who did the work and why, what was the outcome, and how it differed, if at all, from what we have in mind. Each time we read the newspaper, a magazine, professional journal, or book, we are drawing upon experiences of others and information they have developed. Day-to-day discussions with peers and other researchers, engineers, and managers accomplish the same thing.

Perhaps it is unfortunate that most published reports of engineering work stress *positive* accomplishments. None of us cares to admit to failures and disappointments, approaches that did not work, or experiments that showed we were on the wrong track. There is need for more reports detailing projects that were tried that did not work, approaches pursued without success, and just plain honest mistakes. These often have greater value than success stories.

This may be why many workers in technical fields are intrigued with biographies of famous scientists and inventors. It encourages us to know that we are not the only ones who make mistakes and fail, and that if we are perceptive enough we can turn even failures into successes. When asked by reporters how he felt about his numerous failed attempts to perfect the storage battery, Thomas Edison replied that he would admit to no failures, but had learned of several thousand approaches that did not work.[1] Study of failures can be an effective step in avoiding them. They are, in fact, one of the most valuable sources of information for the scientist, engineer, and manager.

Failures = Real-world experience

It is said that products take on a life of their own once they leave the hands of the manufacturer. The same is true for structures, entire plants, and engineered systems. Once the designer's concepts are transformed into tangible materials, parts are made and assembled into systems that occupy real space, and these systems are called upon to perform, they become entities all their own.

Sometimes, their "behavior" and response do not conform to what designers had in mind. This is because they are now out of the laboratory and manufacturing plant and are subject to the multitude of real-world forces and influences. These include the environment and people—directly, as users and operators, and indirectly, as creators of policies, regulations, and laws and founders of social values. Then there are other influences such as economic and political climates on national and global scales.

Not all of these, or their effects, are foreseeable by even the most astute and clairvoyant designer. Some may not have even existed at the time the product was designed, engineered, or constructed. Therefore, we are faced with creating machines, articles, products, structures, and facilities to be used in the fast-changing world afflicted with the decay we mentioned in Chap. 1 and a host of even more hostile forces.

And we must continue to turn out new models having better performance and more efficiency for less cost than what we made last year. It is a tough assignment. To accomplish it, we need all the information on actual performance that we can get. That is, we need reliable feedback. As undesirable as failures are, they provide essential information and experience obtainable in no other way.

Information limits

Failures are an invaluable resource, but they obviously cannot tell us everything we need to know or would like to know. Attempts to distill more information from a failure than it holds can be futile and misleading. Incorrect assessment of information obtained or its misapplication can also be detrimental. And blind acceptance of someone else's diagnosis can worsen your own situation. Intelligent application of failure information demands discriminating evaluation and assurance that it is relevant to the situation at hand.

No two failures are ever exactly alike; therefore, attempts to correlate one outcome with another involving a different set of circumstances, even within the same manufacturing plant and for the same product, must be made cautiously. In studying a given failure, we are observing what happened strictly to *that* plant, machine, device, design, product, or system. The causative network that produced it, as it did and when it did, was unique to that situation, and so were the players, the conditions, flaws and deficiencies, actions and inactions, and all circumstances surrounding and leading up to the culminating event.

Admittedly, we know through experience that there is a certain degree of predictability in the world around us. We develop intuition that can predict the direction things will probably go in if and when they fail. Physics, chemistry, and other sciences apply equally to all without regard to geographic or political boundaries. Iron will rust in Beijing the same as it does in Boston. The same column will buckle whether it is loaded in Miami or Mozambique. Tensile strength of a given alloy is the same anywhere in the world as long as the testing parameters are the same.

Despite such predictability, the underlying causes of failures—circumstances that led up to the event—are generally not predictable or transferable from one case to another. One reason is that human behavior and people's reactions are considerably less predictable than the response of inanimate things.

We cannot expect analysis of someone else's failures to tell us what went wrong in ours. Failure modes may differ as well as causal elements. It is possible for the reported failure *mechanism* (e.g., fatigue, stress corrosion, tensile overload) to be the same but the causation network different. Corrective measures for one might not work for the other. Also, the apparent and reported failure mechanism may not be the actual one.

Therefore, review of failures in someone else's plants, of someone else's products, of someone else's designs, or in someone else's engineered systems may not tell you much about failures in *your* plant, of *your* products, of *your* designs, or in *your* engineered systems. It may offer clues but you cannot always rely on it. How do you know what really led up to that reported failure? Were all details and facts reported? Were they even known? Were interpretations correct?

The goal is avoidance

In failure avoidance we are not solely interested in eliminating recurrences of some particular failure but are attempting to learn how to avoid all failures of that kind. The two are not the same, and different approaches are needed. For example, in the leaking gasoline tank example in Chap. 1, if we concentrate upon that particular occurrence, its failure mechanism, and prevention of recurrences for gasoline tanks in that boat, we might conclude that since it was a corrosion failure a more corrosion-resistant aluminum alloy was called for. Or, someone might conclude that water at the bottom of the tank was the cause and that recurrences would be avoided if the water was eliminated. But would this be a feasible solution for boats operating within a marine environment? What was needed was to remove the copper drawtube that was causing the galvanic reaction in the presence of accumulated water at the low point of the tank and replace it with one of another material not subject to electrochemical reactions.

For failure avoidance we must learn to look beyond the immediate and apparent "cause" of the failure (the mechanism) and identify conditions and circumstances that led up to the corrosion, buckling, or overload. This is where we must start if we are to devise effective avoidance measures.

This does not mean that we should avoid reviewing failure reports or case histories that deal with immediate causes—far from it. But we must consult them for the right reasons. Their principal value lies in the awareness they instill in us of the kinds of failures that can occur under various conditions and circumstances, their "attention-getting" value. It can be useful to know that there have been failures in products or plants similar to ours, even if we have not had any as yet. This awareness should make us and our associates more vigilant. We want to know everything we can about these other failures, as our product or plant may be vulnerable too.

Then there is the educational value of failure reports. Manufacturers frequently design products and operate processes and production lines for operations and materials that are far more complex and potentially hazardous than they realize. Subcontractors especially can find themselves in this situation, as they may not fully appreciate the ultimate purpose for what they are making or the risks inherent in their use.

Engineers of one discipline are not always cognizant of the array of problems that can occur under some conditions in materials or processes they work with. Manufacturers would not bid on some contracts if they fully understood the nature of the work and its liability potential. These people, and others in similar situations, can benefit from reports of incidents, accidents, and failures relevant to their responsibility or activity.

This is probably the dominant reason for the increased popularity of failure analysis and prevention seminars. These are often aimed at a particular industry or product type where problems and failures specific to that segment of industry are discussed in detail. There, case studies of failures directly related to a specific type of product (e.g., semiconductor circuit boards, composite materials, or aging aircraft) are presented with the purpose of training those working in the field about failures and avoidance of failures.

Comprehensive failure analysis, properly done and competently reported and documented, constitutes a strategy of major significance for avoiding failures, particularly so when it is used in conjunction with other techniques described in subsequent chapters. A major disadvantage of failure analyses is that they are retrospective: they deal strictly with the past. They are, as Coleridge viewed history, like a lantern at the stern. They illuminate hazards and pitfalls of the past but offer limited enlightenment for what lies ahead. However, Santayana's often-quoted admonishment on the value of history was never more relevant than in dealing with engineering failures: Those who cannot remember [or choose to ignore] the past are condemned to repeat it. "A page of history is worth a volume of logic," is how one jurist saw it.[2]

Unfortunately, the extrapolation of failure analysis results and conclusions into the future is not simple or straightforward, for the reasons discussed. Yet, the value lies in applying them to new decisions and activities. Consequently, we must develop skills in interpreting and applying findings of past failures to today's designs and decisions.

Anatomy of Proper Failure Analysis

Rationale for investigation

Most of the time, failures of all kinds occur without any systematic or formal post-incident inquiry. This is especially true of minor occurrences, accidents, and mishaps. Usually, formal inquiries with written reports are confined to more serious failures, accidents resulting in injuries or deaths, or property damage and economic loss when the dollar value exceeds some set amount. What constitutes a "reportable" incident (implying an inquiry or investigation of some kind) is largely defined by policies or regulations of some responsible governing body, administrative agency, or office.

The trend for increased litigation stemming from failures has not only broadened the range of types of occurrences that are reportable but has also increased their number. Depending upon the potential for harm and damage, even "near-misses" and infractions of safety regulations are increasingly regarded as reportable incidents. This is so even when no harm, injury, or loss may have occurred. Industries and governmental agencies at all levels have their own sets of rules that dictate when a failure or related incident must be formally investigated. With recent intensified concerns over public safety, health, and environmental damage, there has been a sharp increase in the number and kinds of incidents that are being formally investigated.

We need not debate whether all of this is beneficial and serves a useful purpose. Much of it undoubtedly does. The point is that there are literally tons of incident and accident reports being prepared and countless investigations into failures, mishaps, and other incidents being conducted throughout this country. In view of the number of investigations and incident reports that are being prepared today, we must devote some attention to what constitutes a proper failure investigation, that is, the nature of investigations conducted as part of an effective failure avoidance program, with particular emphasis upon engineering and mechanical failures.

Manufacturers of products with significant liability potential and industrial firms whose activities involve toxic material or high risks of damage are finding it advantageous to organize task forces that can promptly respond to a problem at a moment's notice. These defense or litigation avoidance teams are "packed and ready to go" at any time. Trained in the company's products and processes, they understand the materials and components and the risks associated with them. They can minimize the potential for injury and damage should a chemical be accidentally released, a process become unstable, or a product fail. The incident need not be a failure in the usual sense but may be some occurrence that could have negative effects upon the company. Some examples are product tampering, improper disposal of wastes and associated contamination, and misuse of products.

Team members are well versed in dealing with the media, environmentalists, and other activists. They are knowledgeable about regulations affecting the firm's products and operations and understand corporate policies. If such a team exists, it is usually the one that conducts or coordinates failure investigations and analyses.

Focus and content of failure studies will differ from incident to incident, depending upon the severity and resulting damage, whether the study is conducted voluntarily or is mandated by some regulatory body, and the likelihood of litigation. Generally, it is best to assume that the issues will be litigated and to conduct the study accordingly. Otherwise, essential facts and details may be lost. Critical evidence may become defaced or its chain of custody broken, resulting in damage to some party's interest. It is preferable to err in the direction of too much care to preserve and document evidence than too little.

Should *each* and *every* failure be investigated and analyzed? If we want the benefit of information that it can provide, the failure should be investigated, particularly failures having litigation potential. This does not necessarily mean an exhaustive fragment-by-fragment dissection and detailed reconstruction and analysis, followed by a voluminous report, for every failure. Only enough work need be done to gain a clear and unambiguous picture of what occurred and why; in other words, whatever it takes. Again, if there is a possibility of litigation, much more may need to be done to provide a strong defense and to encourage early and favorable settlement.

Experienced investigators can often make a complete and accurate diagnosis of simple occurrences within a very short time and with minimal supportive laboratory work. A single-page report is frequently all that is needed. The extensiveness of a given study, the time required to conduct it, and sometimes its cost usually are inversely proportional to the experience of the investigator or investigating team. The study and its report should not be any longer or more extensive than necessary for complete understanding of the occurrence and for a reasonably accurate assessment of the causation scenario that is consistent with all the facts and evidence.

Analytical considerations

So far, we have been describing failure *investigations* or inquiries, not failure *analysis*. Many so-called failure analyses are no more than investigations with little analysis being done. Investigations are fact-finding missions. Investigators are concerned with appearance and condition of the failure site, materials and other physical evidence, and documentation of all this information. They are also concerned with people's impressions, accounts of witnesses, and other related information.

Collected information is reviewed, conclusions are drawn, and a report is prepared. Its length and depth of detail usually hinge upon seriousness of the failure, who requested or required the investigation, its purpose, funds available for it, and personal idiosyncracies of the investigator or team. Extensive investigations may include some analysis in the form of reconstructions and exemplar testing (i.e., evaluations of similar or identical articles), but these are usually done to resolve inconsistencies in the findings or to fill an information gap in attempting to understand the event more clearly.

Failure *analysis,* on the other hand, goes beyond the investigation itself; it starts where the investigation leaves off. Analysis explores the *why* and *how,* identifies causation elements and their interrelationships, determines causation sequence, and tries to develop a clear and unambiguous understanding of what led up to the failure event—the last step in the complex sequence that may have begun long before.

For maximum effectiveness as a failure avoidance tool, both investigation and analysis are essential; however, the key to avoidance usually is found within the analysis. Some avoidance clues may be uncovered during the investigation but may not be reliable without analyzing all causative factors. Like investigations, the analysis process may be brief and straightforward or extensive, depending upon the situation being studied and other factors.

The traditional failure investigation of technical matters is usually confined to *physical* aspects: mechanical loads, structural aspects, material response, fracture locations and modes, loss mechanisms, etc. This approach neglects the sticky issues of personal fault, blame, individual responsibility, and accountability. Often, the avoidance is deliberate. There seems to be a strong reluctance on the part of failure investigators to get involved in people-related issues. These are the *who* issues in contrast to the *what* issues, and they relate to decisions, policies, human behavior and response, errors and omissions, responsibilities, and accountability.

Perhaps this is because investigators want to avoid the appearance of a "kangaroo court" or inferences that the study was nothing more than a witch hunt. It may be that technically trained people feel uncomfortable grappling with human behavioral characteristics, personality traits, and motivations that are out of their fields. Whatever the reason, many failure investigators tend to concentrate upon physical matters and are content to let others deal with human deficiencies and personal accountability.

In view of the fact that human errors, deficiencies, and weaknesses, in one form or another, are usually implicated in most failures, it is probably wrong to exclude people factors from most investigations of failures of a technical nature. Certainly, fixing blame does not fall within the province of fact-finding failure investigations. Nevertheless, it is equally wrong not to even consider human factors that may have played a key role in the occurrence. Unless *all* factors are brought out into the open during the investigation (even at the risk of pointing a finger at some individual), the goal of failure avoidance will almost certainly be compromised.

Consequently, traditional failure investigations usually stop with an explanation of the occurrence in physical terms. For example, the pump shaft failed by metal fatigue—period. The implication is that pump shafts are expendable and can be replaced; they "wear out." The objective is to fix it and get the pump back on line as soon as possible. The only concern over the failure might be with regard to the downtime it caused while a replacement shaft was obtained and installed. How should the downtime be prevented in the future? A common response is to simply order additional replacement shafts so that they may be on hand for a short turnaround repair the next time one fails, or, install a spare pump in the line with a bypass valve so that it can be placed into service at a moment's notice.

Such solutions and approach are, unfortunately, not uncommon. The idea is to keep the plant running and not worry about why shafts fail. At first, this may seem practical and effective. But there probably are other components that frequently and prematurely fail, including the bearings of the pump with the failed driveshaft. The common approach can soon become costly in purchases of replacement parts, inventory of spares, repair costs, and downtime, which affects productivity.

Would it not be better to take a little time to learn why the shafts and other components fail frequently and prematurely? Perhaps simple re-alignment of the pump with the drive motor would avoid shaft failures and add life to the bearings as well. Learn why these units were mis-aligned in the first place. Check the units periodically, especially if the pump is critical to the operation and is large, and replacement parts are not readily available.

Effective failure avoidance requires going beyond physical aspects to find out what made the given component fail. Most likely the failure oc-curred because of something someone either did or did not do. Mechanical failures rarely occur because of latent material defects; more frequently they are traceable to human error somewhere along the way. Find it, correct it, and you will be well on the way to avoiding failures, at least of that kind.

Investigative personnel

The investigator's background can influence the direction of the failure investigation and the tone of the findings and even color conclusions and recommendations. Therefore, the choice of the investigator should be made carefully to ensure that the investigation will be complete, thorough, honest, and objective. The investigation is no place for biased interests, prejudices, and subjectivity. The investigator's background and experience should be consistent with the subject matter of the fail-ure and the operations, processes, and applications involved.

It is a demanding task that requires a combination of skills, solid practical experience, and a high degree of perception and intuition. Knowledge of materials properties and behavior is also essential as are common sense, patience, perseverence, logic, and ability to work productively under pressure. Obviously, the investigator should be chosen from among the most competent; failure investigation is not a task to be assigned to losers who have nothing else to do.

For most failures—those of less serious consequences—a single investigator may have to do it all. That investigator must be familiar with all related disciplines or have access to personnel that are. For more extensive investigations, an interdisciplinary team may be needed. The major disciplines involved in the failure should be represented, if not on a full-time basis, then part time or as required. Selection of a failure investigator, team member, or coordinator is a decision that is critical to timely completion of a competent investigation and successful application of its findings to failure avoidance.

Investigation problems and pitfalls

An investigator with the desired traits and capabilities will minimize the potential problems that can arise. However, the more serious ones should be mentioned.

Motivations for conducting failure investigations do not always stem from a desire to minimize recurrences. Some investigations are conducted to absolve a particular party from responsibility or to get someone off the hook. Some seek to implicate others. Some are carried out for promotional, public relations, or propaganda purposes. Then, the media sometimes conduct or sponsor investigations to furnish program material and improve their ratings. Everyone associated in any way with a failure has his or her own interests at heart and his or her own agenda and ax to grind. Usually, no two are the same. Sharply opposed interests are represented at most failure inquiries even before there are well-defined "sides" or lawsuits are filed.

These conditions are not conducive to orderly, objective, and thorough investigations. Most of the time, if the failure has been a serious one involving injuries or significant losses with many interests at stake, it is difficult even to gain access to the failure site and physical evidence. Under these conditions it is easy for the study goals to become compromised.

It may be a truism that failures tend to occur at the worst possible time, but this is often so, as higher priority activities crowd out a failure study and chances for conducting a proper one diminish. For example, a large engineered system is about to be completed. It is important that the contracted completion date is met. During a trial run, a critical component fails, jeopardizing the schedule, and must be replaced. What kind of failure investigation do you suppose is likely to be conducted?

Chances are that it will be superficial and of narrow scope, if one is conducted at all. This is because the principal goal is to restore operation of that unit promptly. The completion schedule here takes precedence over everything else. Will a proper study ever be made of that failed component? Should it be made? What are the risks of recurrences with replacements? What about warranty exposure and contractual effects? Such an incident occurring at a most inopportune time raises many questions. The need to conduct a proper investigation and analysis will probably be shoved aside in the panic over the more urgent requirement to get that unit running again to meet the rapidly approaching deadline.

Perhaps, in the above example, the parties agree that a study should be conducted to ensure some degree of reliability for the replacement. Even that study will be faced with pressures to maintain a narrow focus. It will undoubtedly concentrate solely upon that single part that failed, excluding everything else around it and any in-depth consideration of causal factors. A failure investigation will have been conducted, and some analysis may have been done, but will they offer any significant assistance in avoiding other failures over the long term?

Pressures and stress of schedules, other commitments, and higher priority matters have a way of relegating failure studies to a back burner, especially if they are no longer of immediate urgency. They and the investigators may be accorded center stage when the heat is on, until they provide "an answer" or some form of instant relief for the crisis at hand. But afterward, when the spotlight is gone, it is sometimes difficult to keep the investigation alive, let alone complete it. This certainly is not as it should be, but is mentioned as a pitfall to beware of and avoid, if at all possible.

There are other reasons why failure studies become deflected from their targets. A common one is to save time or money—budgetary restrictions. Another is that a failure study is done merely to fulfill some mandated requirement. Information is collected, facts are documented, a report is prepared; however, it probably will provide no guidance in avoiding recurrences because it was a perfunctory undertaking, perhaps incomplete and unfocused.

Some investigations are conducted in reverse, and this is an easy trap to fall into. The trap may occur at the beginning, but more frequently it is at some intermediate point of the study when there are convincing indications that seem to point to some particular failure mode and cause. These can be mirages but can seem very real, logical, and believable, before all facts are in.

Nevertheless, under pressure from various interests for an early answer, an investigator or team member leaks the observed indications or hypothetical explanations. Before long, they are repeated and with each repetition gather validity. In the absence of other findings, this explanation becomes accepted fact. Under the circumstances, the investigator may be pressured to curtail further in-depth study and "tie things up," which means to concentrate on gathering information and supporting data to reinforce the now-accepted conclusion. The investigation has thus turned around in midstream, and now, with "the explanation" in hand, the task is to make the facts somehow match conclusions.

This inversion of the investigation need not occur deliberately but can take place gradually and subconsciously. The existence of the inversion may come to light only through intensive further study and analysis. Sometimes it is never discovered. This problem can be averted through maintaining tight security on matters involving the incident and its investigation and releasing no opinions or findings until the study has been completed. This is often difficult to do, but must be done to avoid these consequences.

Investigators should also stifle temptations to draw conclusions before all the facts are in, even if they do not disclose them publicly. And it is important to maintain continuous activity during the investigation and avoid gaps of inactivity, such as waiting for laboratory results and other such delays, as these times are ripe for speculation when pressure to release at least "preliminary" findings is greatest.

Conflicting interests

It is usually impracticable for representatives of all parties having interest in the failure to conduct their own investigations, although this is often attempted. Major failures and disasters, with many parties and conflicting interests, are best investigated by neutral parties. These may be independent failure investigators, organizations, or government agencies. They conduct the investigation and their findings are usually made available through appropriate arrangements to the others. This does not prevent the various parties from conducting their own parallel inquiries with emphasis on some specific issue, as insurers, attorneys, designers, contractors, and manufacturers might find it advantageous to do. But it simplifies matters considerably if the principal investigation is conducted by a single neutral party or organization.

For most failures of lesser magnitude, in-house failures for example, corporate management usually calls the shots, and its own staff may conduct the investigation, possibly with help from outside experts or consultants. If a lawsuit is involved, attorneys of both sides often will want to conduct their own studies, employing their own experts. These may not be duplicate studies as their interests and emphases may differ. Sometimes findings of one side's experts will simply be accepted by the other. The dispute in such cases will center about procedural matters or some other legal issue.

Most failure investigations involving multiple parties and interests are not conducted in uncluttered and serene surroundings. The investigator may enter the fray with a well-thought-out agenda and an orderly and logical plan, but may soon become frustrated. Flexibility, determination, and patience are the keys to survival, through perseverance and tenacity, because sooner or later a complete investigation may be conducted, even if not in the planned sequence.

Procedural Detail

The site and its hazards

Failures are unplanned, unexpected, and uncontrolled occurrences. Their extent is sometimes not fully realized until some time later. This means that accident scenes and disaster sites can be dangerous places. Previously solid structures may have been weakened to the point of imminent collapse; apparently stable wreckage and debris may be unstable. The failure may have undermined foundations and supports elsewhere. It may have released toxic substances and filled confined spaces with dangerous concentrations of gases or liquids; combustible substances may be ready to explode. Utility lines may be ruptured, and people who experienced the failure may be injured, disoriented, and shocked and can respond in bizarre and unpredictable ways. As desirable as it may be to examine things promptly in their undisturbed state, it is more important to avoid becoming another casualty.

Risks in the aftermath of failures can persist for some time afterward. Investigators approaching these sites should be suitably equipped, attired in adequate safety gear, and alert to possible hazards of the surroundings. It is easy to become complacent over these matters when preoccupied with gathering information and documenting details.

Checklists

An essential tool in carrying out the investigation and analysis is the checklist. Investigators who are reluctant to prepare and use them because they are viewed as an impediment are reminded that airline and spacecraft crews use them extensively, as do operators of other complex systems.

The checklist must be as detailed and specific to the kind of failure as possible. Generic checklists may be better than none, but will not do the job adequately. Checklists should enumerate specific tasks and the order in which they should be done, although the preferred order may be revised. Examples of checklists prepared for specific failure situations are available.[3–5] They should be consulted and reviewed for general format, degree of detail, and their direct applicability to the types of failure for which they were intended. However, existing checklists prepared for other purposes should be considered strictly as examples and guides for preparing ones for your specific situations.

Checklists should contain a section dealing with collecting and documenting information in the immediate aftermath of an incident. There should also be a section for information on pre-incident conditions and circumstances concerning the equipment and personnel as they were before the incident. A section should be devoted to the incident itself; another to the post-incident situation.

There should be reminder notations at appropriate locations in the checklist of the need to document certain details, such as photographs, witness statements, manuals, sketches, and records. The checklist should also have provisions for collecting physical evidence and related information.

In an effort to relieve some of the paperwork burden and ensure adherence to checklists, software is being developed by Embry-Riddle Aeronautical University to "streamline the aircraft accident investigation process." The concept is for on-site portable computers to display standardized checklists supporting accident report forms, aircraft system diagrams, and/or aircraft specifications and limitations, and also to tie into other information databases containing aircraft, weather, and pilot data. Such developments could be readily transferrable to engineering applications and could facilitate failure investigations.[6]

Investigation plan

Elements and tasks involved in failure investigations are often depicted in tabular form or work diagrams. Their disadvantage is that they must be so generalized that they are not specific enough for actual situations. Most investigations cannot be carried out in the order shown, and as the investigation progresses and facts are uncovered, there can be many needs to retrace earlier steps. The actual sequence may be more like a series of iterative loops that tend to progress downward. Nevertheless, they are useful guides in organizing the necessary steps and investigative sequence.

The elements and sequence in the chart of Fig. 2.1 are presented as typical steps in a failure investigation and analysis. Each investigator should devise his or her own plan tailored to fit the given situation. Procedural details, too, will differ according to the kind of failure, the materials, components, technologies, industry, application, and environments involved, along with many other factors and variables.

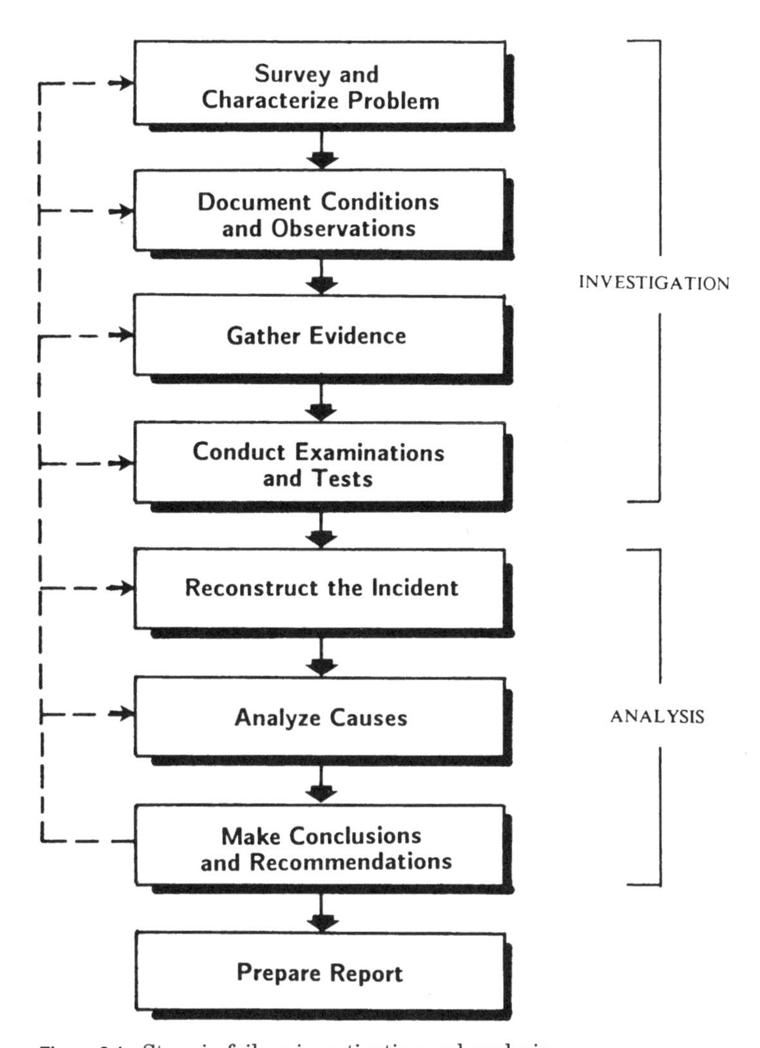

Figure 2.1 Steps in failure investigation and analysis.

The investigation is a learning experience for the investigator, no matter how experienced in conducting investigations he or she may be. Each occurrence is different from all others, although there will be similarities. Learning is usually not a linear and uniformly progressive process. As new facts or skills are learned, they must be digested, assimilated, and compared to previously learned facts and impressions. They must be reconciled with the existing knowledge base before we can take on additional information. Otherwise, there is an informational overload, and our mental apparatus jams and refuses to take on more. Facts or information absorbed must be reinforced through confirming observations or experiences over a period of time before it is fixed within our minds or "learned."

Likewise, in a failure investigation, we must frequently stop, review what has been observed, and reconsider new findings in the light of previous observations. If it "does not compute" it is wise to back up and find out why. It may require reexamination of previous information or reconsideration of the accepted validity of earlier impressions. Sometimes progress requires taking a step or more backward to clarify some puzzling aspect or to reevaluate an implausible hypothesis.

The upper part of the sequence diagram is devoted mostly to investigative matters; the lower to analysis. It is not unusual, during the study, to find that you must return to a previous investigative stage. Perhaps some important detail was overlooked, or the chemical analysis failed to report some critical element. Although Fig. 2.1 indicates iterations originating at the conclusion stage and returning to any previous stage, in actual practice they may originate at any stage. With experience, the investigator becomes proficient in anticipating problems and inconsistencies, in planning the study, and in minimizing iterations.

Defining the problem: Initial survey and characterization

There is always a temptation to dispense with preliminaries and get right to the business of collecting information. Seasoned investigators can be prone to this as they feel confident about being on familiar ground, having seen it all before, and having a fairly good idea of how it all occurred. It is desirable to stifle this impatience, in spite of the fact that time is of the essence in any failure investigation. What happened may be similar in many ways to other previous occurrences. But our interest is not based solely upon whether the failure mechanism was fatigue, creep-rupture, or hydrogen embrittlement, but upon what brought about the situation for some failure mechanism to come into play and do its damage. We must, therefore, look to circumstances, people's actions, and other more subtle elements if we are to find the true culprit. This is essential to successful avoidance of future incidents.

Before ever approaching the failure site and getting on with the investigation, we must have some knowledge of what we will be studying. What had been going on there before the failure? What materials were being handled and used and what machines, equipment, processes, and technology were involved? We must have details on business operations and the industry and its codes, customs, and practices, and more. We should learn everything we can about the operations involved in the failure before we embark on a fact-finding mission at the site.

With a general understanding of what occurred, we are in a better position to devise a plan, checklist, and other investigative tools. We must devise "a wide and flexible net to...pick up complex and unexpected social and technical interactions and side-effects."[7] To do this we must approach the failure scene with a broad perspective that encompasses the physical aspects, the people who were in any way involved, and the surrounding environment.

But before any of this can be done, we need site access permissions and approvals of those officials having jurisdictional authority for the safety and conduct of activities there. This may be a government agency, a regulative body, or their designees. The contacts are necessary to establish the scope of what is permitted and what is not and the conditions governing these activities. This is almost always a consideration for major failures that resulted in significant damage and injuries and major disasters such as explosions, airplane crashes, extensive toxic spills, and train derailments.

Plans and protocol for conducting site inspections and documentation, evidence retrieval, sample removal, and subsequent examination are all usually dictated by the authority having jurisdiction over the incident. This facilitates the investigation, minimizes loss of critical information, and helps protect interests of all parties.[8] Consequently, any investigation and analysis that are done or desired to be done by one of the interested parties must be coordinated through those with authority over, with access to, and with control over activities at the site. In the long run, it is easier to cooperate with these people than it is to try to circumvent them.

It is evident that the various steps in inspecting the failure site, documenting findings and other information, and selecting and examining critical parts are carried out under adverse conditions. The area will be in disarray, it will be dirty and possibly contaminated in some way, and it may be outside and exposed to wind, rain, ice, and snow. There will be some urgency in restoring order and removing wreckage or debris, and there probably will be other investigators jockeying for position to make their own assessments. Items of interest may be buried beneath piles of rubble. These are difficult surroundings in which to attempt to accomplish painstaking examinations and information retrieval. This situation should be kept in mind when preparing for these tasks and when selecting investigative tools, facilities, and apparel for field conditions.

Documenting observations

If there is any likelihood that litigation will result from the incident, documentation should be in accordance with the rules of evidence of the jurisdiction.[9–11] It is advisable to assume that litigation will be involved;

accordingly, documentation should be done by someone knowledgeable in these rules. Documentation includes photographs, witness statements, and other "evidence" generated by the investigation.

When the investigator has surveyed the failure site, has a reasonably good grasp of what occurred, and has identified those items and components that appear to have played a major role, it is time to document general impressions. If at all possible, this should be done before things have been disturbed or attempts have been made to restore operations and order. At this stage, it is important to carry out site inspection and documentation procedures without disturbing anything. Principal tools will be cameras (still and video) and portable audio tape recorders.[12–14] Measuring tapes, rulers, levels, micrometers, magnifying glasses, binoculars, self-contained light sources, etc., also may be needed.[15]

The documentation stage, itself, requires a plan and checklist of its own. Photography should be of professional quality throughout, properly lighted and exposed, and in focus.[16] Supplemental lenses may be required for telephoto (distant) and macro (close-up) work and wide-angle lenses if the area is confined. Color photographs and videotapes provide the most accurate documentation. Use of tripods for both photograph and videotape exposures is recommended.

Views should be in sufficient detail to show the location of key elements with respect to permanent reference points, if possible.[17] This will facilitate accurate reconstruction at some later time. Close-up photographs, especially, should include a measuring scale or ruler, or some familiar object (e.g., a coin) within the field of view for dimensional reference.

Documentation should be extensive and detailed, as the site appearance and the condition of objects and their positions, will never be the same again once things are moved, examined, and the site cleared. The sequence should start with overall views and progress to finer detail. The failed item or area of interest should be photographed from all sides with special care to include warning labels, serial or model numbers, material grade designations, manufacturers' names and logos, and stamped identifying marks and patent numbers. Fractures, scrapes, dents, distortions and misalignments, chips, missing parts of principal objects, and similar views of adjacent items that may have been affected should also be documented.

Extensive failures with widespread damage sometimes require aerial photography for obtaining an adequate perspective. Helicopters are commonly used, as they provide the best platforms, desired vantage points, and required maneuverability. If it is too late to capture important features, local television studios, newspaper offices, or amateur photographers in the area may have photographed or taped the accident in progress or soon after. These sources for documentation of major failures should not be overlooked.

Witness accounts also must be documented. The most effective way is using a videotape or audio tape recorder. The witness should be identified on camera or on tape along with the place, time, date, and interviewer's name and business affiliation. Checklists for questions are essential to ensure coverage of all relevant points. Sample interrogatories, obtainable from legal texts and practice aids, and other sources that cover a wide range of situations are useful guides.[18,19] Written statements, signed and dated, whether in the handwriting of the witness or typed, may also be obtained.[20]

Other documentation that might be obtained as appropriate to the failure occurrence are

Sketches

Maps

Floorplans

Product and process flowsheets

Wiring and assembly diagrams

Test data

Inspection records

Copies of operating procedures and user manuals

Specifications

Safety procedures and records

Purchase orders

Applicable codes

Standards and regulations

Repair and maintenance records

Shop orders

Manufacturer's directives

Personnel files and employee time cards

Sales literature

Brochures

Advertising

Patents and publications

Even this list is not exhaustive. Other forms or documents may be relevant and will assist the study.

Information on unfailed duplicate products or equipment that may have seen similar service or may have been operated within the same plant can also be useful and should be documented. Even operating histories for identical or similar equipment or operations at other corporate facilities at other locations may be helpful in analyzing the failure.

Gathering evidence

This step is simply the collection of tangible materials and other documented information noted above. The range of what might be included is extremely wide; it can be anything that will shed light upon the investigation. The goal is to obtain enough information and whatever materials will be required to provide an unambiguous account of the incident. In addition to the documents noted, it might also include samples of materials involved in the failure, fragments of damaged equipment, or the entire item or assembly that is believed to have been directly involved, if it is possible to obtain it.

Deciding what is significant and what is not requires considerable insight into the occurrence and what brought it about—the reason for the "homework" and preparation mentioned earlier. The investigator should be able to distinguish between damage that triggered the occurrence and damage that was the result of it. Focusing upon secondary effects and similar blind alleys can be wasteful and can lead to a misleading or inconclusive outcome. This is a major pitfall *throughout* the investigation, not merely at this stage, and must be continually resisted.

Failures of massive equipment, structures, and other large systems present problems in recovering evidence. Photographs are an essential first step, but more is usually needed. It is possible to cut out certain cross sections of interest for laboratory examination, but there may be objections to this, especially if there are many parties having conflicting interests. If done at all, it is usually through consensus agreements coordinated by the site authority, or even court orders. These approvals, let alone the actual sampling and examination, can take a lot of time. Meanwhile, there is always the possibility that critical items will be scrapped or otherwise lost before their importance is realized. Field techniques can sometimes assist in retrieving information without the need for sectioning and removal of actual hardware.

Impressions of fracture surfaces and similar areas of interest, such as welds, can be cast in silicone rubber and positive replicas in a suitable resin prepared from them.[21–23] For microstructures, selected surfaces can be metallographically prepared in the field and portable microscopes used to make photomicrographs without removing or otherwise affecting the part.[24,25]

Hardness measurements can also be made and chemical composition can be determined in the field by x-ray fluorescence methods using portable equipment. Even scanning electron microscopy (SEM) examinations are possible using compact instruments designed for field application. In the absence of such facilities, small samples of surface films, residues, corrosion products, operating fluids, and metal fragments can be collected for later laboratory examinations.

For reference purposes, it may be helpful to determine if equipment similar or identical to that involved in the failure, but still operational, is available for inspection and study. A good deal of information may be obtained from this and may not be as subject to examination or sampling restrictions as those directly involved in the incident. Condition-monitoring techniques described in later sections of the book may be applicable in these examinations. Samples of unaffected production runs or batches can be useful in identifying specification deviations.

Care is needed in handling samples that are inspected at the site or are retrieved for laboratory examination and transported elsewhere. Critical surfaces must be preserved and contamination avoided. It is important to photograph each part or component that is sectioned or removed in order to document its orientation and origin and preserve its identity.

Each step should be clearly documented and should include time and date, who sectioned or removed the piece, the method used, and other pertinent information that will offer a clear historical record. A written record of the entire disposition history of each sample and part removed for study should be maintained as others take custody of it for their own testing and evaluation and pass it along to other recipients. If the part is or becomes evidence in a lawsuit, it can be necessary to demonstrate an unbroken chain of custody.

Conducting examinations

Whenever possible, it is advantageous to conduct examinations away from the failure site. Even if affected parts or equipment are too large to be taken to the laboratory, it is desirable to protect them from weather, dust, and other contamination. Makeshift shelters can be erected over and around materials being studied. Or, if feasible, move them to adjacent buildings that can serve as temporary laboratory facilities. Otherwise, samples, evidence, and documented materials should be taken to appropriate facilities where studies can be performed in suitable surroundings. As in previous steps, the examination sequence should be planned ahead of time to ensure that tests and examinations are conducted in the right order.

It can be useful to assemble fragments and wreckage of larger failures in roughly the same configurations and orientations that existed prior to the incident. This arrangement can assist in identifying fracture origins and other failure initiating points. Such layouts should be well documented from every angle, as was done at the original failure site.

Kinds of examinations conducted will differ considerably with the nature of the failure, components involved, materials of construction, and many other factors. The purpose of examinations is to obtain further information—enough to allow investigators to gain a full understanding of what occurred and why. This stage, the final one in the investigative portion of the sequence (see Fig. 2.1), is the source of the input for subsequent analysis. Results of examinations conducted here furnish pieces of the puzzle that often are crucial in pinpointing the cause. The examinations should enable investigators to reconstruct the incident and establish the causative pattern that led up to it.

The best scheme to follow in determining what to examine and how to proceed is to attempt a preliminary reconstruction—a mental or paper exercise. This may be one or more plausible failure scenarios that incorporate all findings and information available so far. These should be held in strict confidence by investigators and not disclosed to outsiders. Several scenarios may be possible at this stage. It does not matter, as their purpose here is simply to identify information gaps. When this is done, examinations or tests can be devised to furnish the missing information. There never is enough time and money to examine everything and we must narrow the field of choices.

Throughout this stage, examinations should progress from the general to the specific. In other words, use sophisticated diagnostic methodologies only when less elaborate techniques fail to provide answers; use high magnifications only after lower magnifications reveal areas of principal interest. It is easy to get lost and waste valuable time in starting off with high-resolution examinations of some general feature of interest that may not be relevant. Initial examination should be confined to visual inspections. For now, ignore high-technology diagnostic gear.

Preliminary examinations will, of course, raise questions and uncover the need for more elaborate techniques. Steps may then be taken to answer these questions using whatever facilities are available *to clarify these issues.* Investigators must be self-disciplined to resist impulses to explore aimlessly. The examination sequence should be dictated by the findings as examinations progress from a low level of detail to a higher one in search of answers to questions that arise.

Discussions so far have assumed that the failure being investigated is of the traditional variety, in which a manufactured article, engineered system, or industrial process went wrong, with resulting physical damage, broken parts, debris, and, perhaps, personal injury. By the broadened definition of failure presented in Chap. 1, many kinds of failure can occur without tangible or physical effects. It should be mentioned that the procedure for investigating failures of all types, with or without physical damage, does not differ significantly from one to another failure. Investigative steps and their sequence should not change; the same format can apply for all failures. This holds true for analysis as well. Investigative details and examination techniques may change but the approach and philosophy remain the same.

In this context, it is worth mentioning that all mechanical failures do not require a full complement of traditional tests for proper analysis. That is, intensive examinations may not be needed to determine failure mechanisms or causal sequence. This does not say that investigation and analysis are not required, only that tasks usually associated with failure analysis may not be necessary every time. This is why it is important to maintain a clear perspective throughout the procedure. Do not become so immersed in the smallest details that the principal cause is overlooked.

Notwithstanding the occasional possibility of correctly assessing the failure without running a rigorous program of tests, it is often necessary to conduct extensive examinations of failed parts to characterize the failure accurately and to determine what caused it. For mechanical failures, materials testing laboratories are often involved and metallurgical evaluation procedures are commonly necessary. Table 2.1 is a listing of examinations and tests conducted during studies of mechanical failures. The list seems extensive but is not complete.[26–28] There are numerous variations for some tests. Corrosion and weld tests are examples, and mechanical tests may be conducted over a range of temperatures and environments to simulate failure conditions.

Tests are conducted during failure investigations to verify properties or composition, to establish code compliance or its absence, to determine material condition or quality, to evaluate responses under known and controlled conditions, and to characterize some aspect of the material or its response during the failure. There are standardized procedures for most tests and examinations. These are important to observe to minimize errors and scatter and to serve as a basis for test comparisons.[29] Sometimes, there is insufficient material for conducting standard tests, and techniques must be improvised or alternative techniques employed using subsize specimens.

In addition to tests conducted on fragments and other materials directly associated with the failure incident, it may be desirable to conduct tests of ongoing processes and operations. Tests described in subsequent chapters that deal with monitoring the integrity of opera-

TABLE 2.1 Examination and Test Methods Used in Studying Mechanical Failures

Nondestructive inspection	Chemical analysis
Magnetic particle	Wet chemistry
Liquid penetrant	Emission spectroscopy
Eddy current	Atomic absorption spectroscopy (AAS)
Ultrasonic	X-ray diffraction (XRD)
Radiographic	X-ray fluorescence (XRF)
Acoustic emission	Inert gas fusion
	Gas chromatography
Surface characterization	Combustion + gravimetric
Optical microscopy	Spot tests
Scanning electron microscopy (SEM)	Auger electron spectroscopy (AES)
Energy-dispersive spectroscopy (EDS)	Secondary ion mass spectroscopy (SIMS)
X-ray mapping	Electron-microprobe analysis (EMP)
Mechanical tests	X-ray photoelectron spectroscopy (XPS)
Tensile	Scanning Auger microprobe (SAM)
Torsion	Transmission electron microscopy (TEM)
Hardness	Visual inspection
Impact	
Drop weight	Dimensional verification
Dynamic tear	Stress analysis
Fracture toughness	
Bend	Exemplar tests
Fatigue	Simulated service tests
Creep rupture	
Stress rupture	Corrosion tests
Wear	Welding tests
Formability	
	Differential thermal analysis

tional equipment may be adaptable for this purpose. These tests are useful in providing information to assist in developing failure avoidance strategies, particularly if their results are correlated with findings and conclusions of the failure investigation and analysis.

Examinations are conducted to shed light upon the failure incident, and it is not always possible during the early stages to anticipate what will be needed. Additional samples or information are often required. Perhaps the wrong samples were obtained. If so, the failure site or materials repository must be revisited and attempts made to locate the desired material. Photographs taken earlier can be helpful in locating the right piece. If not too much time has elapsed, what is needed may still be obtainable.

Incident reconstruction

As preliminary reconstructions were done during the previous stage, and examinations carried out to fill information gaps and help narrow the number of possible failure scenarios, a dominant scenario may have emerged. This is simply a hypothetical but plausible sequence of circumstances, conditions, decisions, and events that could have led up to the failure and made it occur, and where *all* causative elements converge.

A reconstructed scenario is the product of all preceding steps, and the remaining steps of the investigation depend heavily upon it. It is essential, then, that this scenario accounts for all the facts and is consistent with all hard and indisputable evidence. There are bound to be minor discrepancies but the objective is a best fit. Simply let the information uncovered during the investigation speak for itself. Avoid speculation, rationalization, intuition, and inferences based upon lack of information.

Extensive reviews of engineering failures reveal that the vast majority do not occur by complex scientific phenomena. Most are the result of faulty judgment. Defective materials and mathematical errors play a minor role. The causal network may be unique but the physical occurrence is almost always explainable with familiar technology. The correct diagnosis is usually the most obvious one.

Under the right combination of conditions and circumstances everything will fail and will do so by the most simple and direct route consistent with those circumstances and conditions. Reconstructions, therefore, should be based upon the most fundamental and straightforward explanation. Complex failure mechanisms and intricate scientific hypotheses should be rejected, or at least deferred, until demanded by facts and evidence.

At this stage there still may be considerable uncertainty over which of several possible scenarios is the correct one. Further testing may be needed for clarification. The task boils down to identifying elements of uncertainty and devising ways to resolve them. It may mean testing a full-scale mockup, calculating stress levels using standard formulas, perhaps conducting finite-element stress analysis to obtain a better understanding of loading effects, or running tests in environments under which the failure occurred.

Tests conducted on duplicates of the failed component (exemplars) or on samples of unaffected material from the failed item may be helpful. Their test response is compared with characteristics of the failed part and checked for similarities. It is important during these tests, as throughout the study, for investigators to remain objective and unbiased. This can become difficult when there is a temptation to rationalize the test results and indications that do not support some previously conceived hypothesis. Open-mindedness is the key to keeping on the track and arriving at the correct diagnosis.

A good understanding of properties and behavior of the materials involved, and how they respond under various situations, is indispensable in resolving difficulties. For example, to explain repeated failure of high-strength steel bolts, knowledge of their susceptibility to hydrogen embrittlement and catastrophic fracture while in direct contact with an aluminum flange in a moist environment was essential. Efforts by maintenance personnel to replace failed bolts with increasingly stronger ones had merely shortened the time to failure because of the increasing susceptibility to this kind of failure in higher strength steels.

In the gasoline tank example in Chap. 1, attempts to provide extra-reinforced welds at the tank corner after leaks had occurred there did not prevent recurring leaks because conditions responsible for localized corrosive attack from within the tank had not been understood or eliminated. Also, unexpectedly large increases in gas flow rates through minute leak paths under very-high-pressure differentials played a major role in the explosion of the stainless steel canister of the other example in Chap. 1. A canister leak was not considered a strong possibility at first because the vessel had been carefully checked earlier with state-of-the-art leak-detection equipment.

Unusually perceptive foresight and a high level of materials expertise would have been necessary ahead of time to prevent some of these materials failures—attributes not possessed by many engineers, let alone by technicians and maintenance workers. Of course, once the failures have occurred it is relatively easy to reconstruct the puzzle. If investigators find that a coherent explanation requires assistance of other disciplines, they should promptly call experts in, without hesitation or apology. Competent, correct, and complete understanding of the real causes is critical in avoiding future occurrences.

Causation analysis

Failure reconstruction in the previous stage tends to focus upon physical and tangible factors that made the failure occur. Perhaps this is because it is easier to visualize and conceptualize occurrences involving objects and technologies or because these factors are closer to the background and experience of engineering failure investigators. Whatever the reason, many failure analyses stop there and do not devote much attention to human behavior and related causative factors. But, in any true failure *analysis,* these must be included. The job cannot end with an explanation of physical factors if the purpose of the study is to avoid recurrences and future incidents of similar failures.

It is evident from the preceding section that some prior analysis of causes is necessary for accurate reconstruction. The division of activities is, admittedly, somewhat arbitrary, but useful for discussion purposes. In this step the emphasis is more upon nonphysical aspects and how they interrelate.

Reconstruction and analysis of relatively simple failure incidents may be carried out mentally. We simply reason it through without conscious awareness of the stepwise analytical process or logic being employed—and it works. From conclusions derived in this way it is usually not difficult to also come up with recommendations for avoiding recurrences. But as failures become more complex, with several causative elements and situations that interact, informal analysis may not be sufficient.

There are no hard and fast rules as to when formal approaches are required. The analyst readily recognizes when the volume of information and facts become overwhelming. Reluctance to admit this, or dogged determination to wing it and press on in spite of it all, can cause loss of important details, causative links to be overlooked, and incomplete conclusions. Even during reconstruction stages, it is often helpful to use diagnostic aids (diagrams, charts, and other graphic means) to assist the analyst in comprehending the entire picture. The benefits are twofold: they provide analytical strategy plus give explicit notation for documentation, discussion, and review.

Method and format are unimportant as long as they accommodate essential details and help the investigator-analyst reach correct conclusions. Various logic diagrams and formats have been developed for dealing with complex systems having many interacting elements. Some of these are discussed in more detail in the following chapter, but it will be helpful to briefly mention them here.

Event-tree or *fault-tree* diagrams are employed mostly in assessing susceptibility to failures and other problems *before* occurrences; however, they can be adapted for analyzing failures. For relatively minor incidents, the tasks of constructing rigorous logic diagrams can be an impediment. Simplified formats can often serve the purpose. They not only help the analyst visualize causal links and overall effects but also impart a thorough understanding of the occurrence.

Fault trees, as usually constructed, place the failure event or undesired effect at the top of the diagram. Then conditions or events are identified (in lower branches) that could bring about the unwanted result. All relevant contributors are included, ranging from those having the most direct effect to indirect causes. Their location in the network or distance from the "main event" does not imply lesser importance, merely their logical position in the progression. In fact, key elements in the causal network often lie at these lower levels.

In its primary use as an avoidance tool, the practice has been to analyze the situation from the causative elements upward to the failure event. This is because effective avoidance must start on lower branches.

Figure 2.2 depicts a simplified causation (fault tree) model for the leaking gasoline tank example of Chap. 1. It provides an explicit representation of the situation and facilitates development of recommenda-

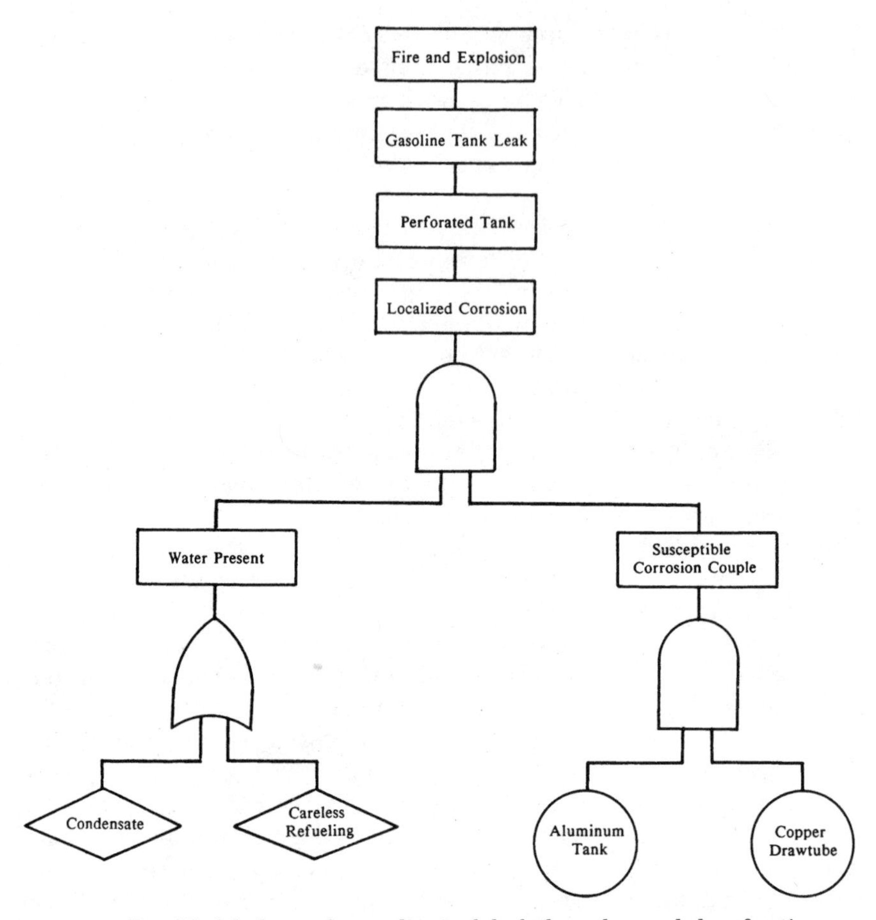

Figure 2.2 Simplified fault tree for gasoline tank leak through corroded perforation.

tions and alternative actions to avoid recurrences. In addition, it ensures that all elements are accounted for. As failures become more complex with many additional causative elements, the need for tangible models and more detailed and rigorous versions increases significantly.

The more typical fault-tree models used in failure avoidance (in contrast to failure *analysis*) consider a broader range of causative elements, as their purpose is to identify potential problems and susceptibilities. Accordingly, they contain more detail. For comparison purposes, Fig. 3.3 is a more comprehensive fault tree for the same example. (The reader is referred to the following chapter for details on fault-tree construction and explanation of the symbols used.) A number of faults could have produced the unwanted top event and some have been included on different branches within this model. Note how this expanded format provides more information on avoidance as it contemplates a broader range of possibilities.

During this stage of analysis it is desirable to consider how the extent of damage and other consequences of the failure might have been averted or diminished. Some major failures might have been only minor incidents if specific, very simple, and, often, almost trivial actions had been taken at the right time, or if some essential ingredient for the occurrence had been lacking. These "intervenors" should be identified, as they are invaluable input for making avoidance recommendations. Therefore, any additional effort that can identify elements that contributed significantly—either to the occurrence of the incident or to diminishing its extent—is effort well spent.

Conclusions and recommendations

Conclusions relate to the failure incident itself and the set of causative elements that brought about its physical or tangible effects. Previous steps should have provided all information needed for sound and valid conclusions that are objective and wholly supportable by the evidence. Formulating them involves considerably more than simply reducing the findings to a series of succinct phrases that summarize results. This is the place where all issues are brought together, where loose ends are tied, and where discrepancies are reconciled. Yet, the conclusions must extend no further than the findings warrant and they must be clear and unambiguous.

Unfortunately, too often those directly concerned with failure occurrences and their consequences are interested solely in conclusions of the investigation. Investigative details, procedures, and evaluation techniques become incidental, as conclusions constitute the "bottom line" for many people. This is what they have been waiting for and probably paying for. Conclusions are considered *the* product of the failure investigation; they are what the media report and what will be quoted during litigation. They may be the sole basis for subsequent corrective action by management or other authorities having jurisdiction over the incident. For these reasons, care must be taken in their preparation.

The task of drawing conclusions is not so difficult if results of the study are straightforward and clear-cut, with few complications. But not all failure analyses are of this kind. Essential facts are sometimes lost and irretrievable. Key evidence can be obliterated or unavailable for study. Intervening conditions between the time of failure and its investigation can make observations questionable. The trail of causation is obscured by passing time and circumstances. These are but a few of the many reasons for inconclusive results. If such is the case, the conclusions section of the report is not the place to attempt to correct the situation, to rationalize, or to speculate.

Recommendations should reflect the conclusions and be consistent with factual findings. This is the focal point of the avoidance value of the entire investigation and analysis. It is a listing of avoidance techniques for future implementation. The recommendations should be specific and clearly supported by analytical results. It is useless to recommend measures that are impracticable or not feasible, regardless of how effective they might seem to investigators. Discontinuance of an essential activity or use of materials having no viable substitutes are not reasonable recommendations. Nonobvious or questionably high risks associated with operations involved should be mentioned, however, even if no practical alternatives exist at the moment.

In developing recommendations, it is useful to recall that discussions of avoidance must include considerations of preexisting conditions and the complete causal network. Effective avoidance measures cannot be founded upon considerations of the triggering failure event alone. Consequently, the best recommendations are obtained directly from the causation analysis.

Referring, for example, to Fig. 2.2, a recommendation to avoid explosions on pleasure boats from gasoline leaks would start with measures that would correct the listed faults. If the objective was confined to avoiding recurrences of the same kind of failure that was experienced, these diagrams show that localized corrosion could be avoided by eliminating the presence of residual water within the tank or eliminating the troublesome dissimilar metal couple.

The latter could be done by substituting a different material for the tank or for the drawtube. From a practical standpoint, elimination of residual water within a fuel tank in a boat in continuous operation in a marine environment probably is an impracticable recommendation. But replacement of the drawtube with another material that would not generate an electrochemical reaction with the tank would be feasible. That should arrest the corrosive action and thereby avoid recurrences.

For the broader objective of avoiding gasoline leaks on boats, the expanded fault tree of Fig. 3.3 offers additional guidance. It addresses the potential problem in using a susceptible metal combination in the presence of moisture, including accumulated water within the tank. It also addresses several other potential problems: ruptured fuel lines, leaky fittings, fatigue, defects, and other damage. There are other possibilities that are not included. Broad recommendations for avoiding gasoline leaks on boats and their explosion potential may be developed from the fault tree of Fig. 3.3. Obviously, they would be more extensive than those from the single fault chain depicted in Fig. 2.2, which was confined to conditions specific to the observed failure.

For situations where analysis reveals significant and perhaps unacceptable risks associated with continued operation of some process or activity, the analysis and its graphic aids can often suggest alternatives. For example, if it were considered that risks of gasoline leaks and their explosion potential on pleasure boats were unacceptable, even when all precautions are taken to avoid identified faults, a less volatile fuel and an appropriate replacement engine may be considered.

Generic conclusions may not contribute to the practical goal of avoiding recurrences, although they are commonly reported. That is, true causes are often identified for which there are no practicable avoidance recommendations. An example is in attributing the cause of an airplane crash to *pilot error*. Pilot error or errors may, in fact, have been the cause of the crash, but how can this be avoided in the future? More training is often a suggested answer.

But is this a reasonable answer? Certainly, airplane pilots must be well trained, but what of the pilot error conclusion when all flight crew members had just completed refresher courses and were known to have been well-trained veterans with years of safe flight operation? Sooner or later, since pilots are human beings, they will make mistakes that may have disastrous consequences. Do we avoid future airplane crashes by ceasing all future air travel? Do we develop pilotless airplanes? These are impractical solutions.

It has been suggested that if even trained professional pilots are prone to committing fateful errors, perhaps the solution lies in learning how to prevent airplane disasters even if the pilot does make an error. Better yet, accept the fact that pilot errors are inevitable, and devise aircraft systems that somehow can tolerate the error, override it, ignore it, or compensate for it, and thereby avert a crash. This may not be altogether immediately or readily achievable, although progress has been made in recent years along this line, but this would be true failure avoidance.

The airplane pilot example is one of many but, because of its widespread prominence by media treatment of airplane crashes, it served to illustrate the problem. If, when faced with making recommendations for failure avoidance, we find we are up against the impenetrable wall of human frailties, it might be time to look to approaches for accommodating those frailties and resulting inevitable errors and omissions. For the given situation, there may be no other reasonable alternative for failure avoidance.

Reports and implementation

It has been said that they who know but cannot express what they know might as well be ignorant. The most capable analyst with the keenest insights into the intricacies of failure causation will serve little useful purpose if that analyst cannot clearly communicate these insights to others.

The report is the vehicle that gives substance and expression to findings, conclusions, and recommendations. It either stimulates the reader to action in implementing its recommendations or lulls the reader into filing the report "for future reference." The report is the tool for making things happen. If it is disorganized and its explanations are obscure, resources that were committed to the failure study are wasted. As a result, nothing will happen and similar or worse failures will continue.

Use of professional report writers to get the job done is not a good approach for reporting results of failure analysis. Investigators or team members have lived through the investigation firsthand, seen and handled fragments and evidence, planned the study and its steps, made decisions based upon their observations, devised examination techniques and test methods, and personally heard witness accounts and consulted with experts of other disciplines. They have reconstructed the incident and formulated conclusions and recommendations. These are personal encounters, personal experiences, and products of personal creativity.

Writers who did not participate in these activities, decisions, and experiences cannot hope to fully comprehend the intricacies of the investigation, interrelationships of the causative elements, and the significance of the findings. Reports prepared by surrogate writers, therefore, can be deficient in critical detail and emphasis. It is best if the written report, as with each step of the study, is the work of individuals who personally conducted it.

The report, as with all effective communication, must be directed to receptive readers. Generally, but not always, failure investigators will not be the ones who implement recommendations. Since implementation of recommendations for avoidance may require significant allocations of resources, operational modifications, and organizational changes, upper management will probably be involved. Therefore, unless the report is received, read, and understood by those at this level with appropriate responsibility and authority, effective implementation probably will not occur.

It is possible that the recommendations may not be amenable to implementation, or circumstances may prompt alternative action. This is not a responsibility of the investigators, but investigators owe sponsors and clients a clear account of the occurrence and realistic and reasonable recommendations consistent with the policies and business objectives of the organization.

To ensure a successful outcome, it is worthwhile for the investigator or team leader, throughout the study, to maintain frequent contact and an open communication channel with those who initiated the study and are responsible for implementing recommendations. Guidance from this direction should be actively sought from inception of the study through completion of the final report. Unless this is done, it is probable that the study's effectiveness will become blunted, its investigators will lose touch with management, and their recommendations will be out of step with corporate policy or agency goals.

Reaping Benefits from Adversity

In the engineering world, failures and accidents have often opened the door to progress. Scientific pioneers and inventors have commonly attributed their successes to earlier failures. Sir Humphrey Davy openly admitted, "The most important of my discoveries have been suggested to me by my failures." As already noted, Thomas Edison viewed his thousands of failed experiments as but steps along the path to success in showing him what did not work. A philosopher echoing this outlook said that "good judgment comes from experience; experience comes from poor judgment." Henry Petroski, in his excellent book on the role of failure in successful design, remarks that

> Success may be grand, but disappointment can often teach us more. It is for this reason that hardly a history can be written that does not include the classic blunders, which more often than not signal new beginnings and new triumphs.[30]

Every learning path is marked by a succession of stumbles and reversals. Virtually all technological breakthroughs have been born out of failure, disappointment, and, often, tragedy. Exploding boilers of early steam-powered devices and the deaths they caused prompted investigations that paved the way to the standards organizations we have today.[31] The more recent catastrophic breakups of Liberty ships led to our improved understanding of brittle fractures.[32] The examples are numerous.

Despite advancements and technological achievement, failures still occur, some of them in catastrophic proportions. As dreadful and unfortunate as these incidents are, they are opportunities for progress and for readjusting our focus on how we do certain things. We must learn to make the most of these opportunities. Many have paid, and are still paying, a heavy price in tuition for our education in these matters and we cannot afford to ignore these courses when they are offered.

Catastrophic failures are more widely known because of the publicity they receive. Frequently, they are spectacular public displays. Curiosity seekers flock to them and visually oriented news media know that they increase viewing audiences. Investigation of the failures may receive some publicity but not as much as the occurrence. By the time causes are identified and conclusions are reached, the incident is history and investigation results are lost on some back page of the paper, if reported at all.

Less spectacular failures are not newsworthy and their accounts do not reach wide audiences. In-plant failures involving manufacturing or process operations frequently receive little outside notice and even insiders may not be aware of them. None of us wants publicity for our failures, as they imply carelessness or ineptness, or that we have been in some way deficient. Since there is something positive to be learned from virtually all failures, it is unfortunate that these benefits are withheld from the vast majority.

There is probably nothing much we can do about most of this. But, in our engineering world, we should be able to recover values that lie buried within the growing mountain of failure reports in our industries. The engineering profession has, admittedly, been lax in recognizing the need for failure information feedback to its members and in doing something constructive about it.

By contrast, the medical profession devotes an entire branch, pathology, to failures of the human body. It is a well-established discipline with an extensive database readily accessible by practitioners. Similarly, attorneys and liability insurers who, like engineers, have interests in accidents and failures and their causation, have sophisticated and well-developed computerized databases and indexing systems offering failure incident information literally at their fingertips and at modest cost.

Engineers need an equivalent system. In a brief but thought-provoking piece on this issue, Neal FitzSimons concludes that

> The greatest engineering failure of all is the failure of an entire profession to equip itself with a system by which it can learn from both its failures and its successes.[33]

While steps have been taken during recent years by the chemical process industry to correct this lack,[34] much more remains to be done if we are to use failure analysis effectively as an avoidance strategy. One way to start is by familiarizing ourselves with databases that do exist in our own as well as other professions. Many others are directly applicable to engineering situations and can be of immediate assistance in investigating and analyzing failures. Besides, they contain useful guidance on investigative techniques and procedures.

References

1. Brian G. Mulconrey, "Edison's Greatest Invention," *The Wall Street Journal,* July 13, 1992, p. A14, col. 3.
2. O.W. Holmes, in N.Y. Trust v. Eisner, 256 U.S. 345, 349 (1921).
3. Charles E. Witherell, *How To Avoid Products Liability Lawsuits and Damages,* Noyes, Park Ridge, NJ, 1985, pp. 285–292.
4. Edward M. Swartz, *Proof of Product Defect,* Lawyers Co-Operative, Rochester, NY, 1985, Secs. 4:8, 6:15, 7:20, 8:17, 12:12, 14:10, 16:6.
5. *American Law of Products Liability,* 3d ed., "Practice Aids," Lawyers Co-Operative, Rochester, NY, 1987, supplemented, Secs. 1–30.
6. "A Better Way to Investigate Accidents?," *General Aviation News & Flyer,* second September issue, 1992, p. A7, col. 1.
7. Victor Bignell, G. Peters, and C. Pym, *Catastrophic Failures,* second printing (revised), Open University Press, New York, 1978.
8. "Standard Practice for Examining and Testing Items That Are or May Become Involved in Products Liability Litigation," ASTM E860, *Annual Book of ASTM Standards,* American Society for Testing and Materials, Philadelphia, PA, 1993.
9. M. Graham, *Handbook of Federal Evidence,* West, St. Paul, MN, 1981.
10. Paul M. Rothstein, *Evidence—State and Federal Rules,* 2d ed., West, St. Paul, MN, 1981.
11. Edward W. Cleary (ed.), *McCormick's Hornbook on Evidence,* 2d ed., West, St. Paul, MN, 1972, with supplements.
12. Gale E. Spring, "Proof Positive," *Industrial Photography,* May 1988, pp. 29–32.
13. Anthony M. Golec, *Techniques of Legal Investigation,* Charles C. Thomas, Springfield, IL, 1976, pp. 91–105.
14. Jack Behrend, "Videotape or Film As Evidence," *Case & Comment,* March-April 1981, pp. 29–32.
15. Donald O. Cox and George E. Moller, "Practical Aspects of the Engineering Problem Investigation," *Fracture and Failure: Analysis, Mechanisms, and Applications,* Proceedings of the American Society for Metals Fracture and Failure Sessions at the 1980 WESTEC, 17–20 March 1980, American Society for Metals, Metals Park, OH, pp. 119–128.
16. John E. Fletcher, "Basic Requirements of Photographs to Become Admissible Evidence," *Functional Photography,* January-February 1987, pp. 16–18.
17. Jack Whitnall, Kimberly Millen-Playter, and Francis H. Moffitt, "The Reverse Projection Technique in Forensic Photogrammetry," *Functional Photography,* January-February 1988, pp. 32–38.
18. Douglas Danner, *Pattern Discovery: Products Liability,* Lawyers Co-Operative, Rochester, NY, 1985, with supplements.
19. Reference 5, op. cit., Secs. 1–700.
20. Anthony M. Golec, op. cit., Ref. 13, pp. 53–90.
21. P.L. Threadgill and P.N. Hone, "Fracture Replicas for Failure Investigations," *Research Bulletin,* The Welding Institute, Cambridge, England, December 1983, Vol. 24, No. 12, pp. 395–398.
22. Robert F. Smith, "Micro Replication," *Industrial Photography,* May 1989, pp. 32–33, 40.
23. Harry B. Hollander, *Plastics For Artists and Craftsmen,* Watson-Guptill, New York, 1972, pp. 169–181.
24. Hugh Crowder, "A Portable Metallographic Laboratory Suited for On-Site Failure Analysis," *Metals Handbook,* 8th ed., Vol. 10, *Failure Analysis and Prevention,* American Society for Metals, Metals Park, OH, 1975, pp. 26–29.
25. Eric V. Sullivan, "Field Metallography Equipment and Techniques," *Micro-structural Science* (M.E. Blum, P.M. French, R.M. Middleton, and G.F. VanderVoort, eds.), American Society for Metals, Metals Park, OH, 1987, Vol. 15, pp. 3–11.
26. *Metals Handbook,* 9th ed., Vol. 8, *Mechanical Testing,* ASM International, Materials Park, OH, June 1985.
27. George E. Dieter, *Mechanical Metallurgy,* 2d ed., McGraw-Hill, New York, 1976, pp. 329–527.

28. Hobart H. Willard, Lynne L. Merritt, Jr., John A. Dean, and Frank A. Settle, Jr., *Instrumental Methods of Analysis,* 6th ed., Van Nostrand, New York, 1981.
29. Patricia L. Ricci, "Standards Sources and Resources," *ASTM Standardization News,* June 1990, pp. 54–59.
30. Henry Petroski, *To Engineer Is Human—The Role of Failure In Successful Design,* Vintage, New York, 1992, p. 9.
31. Sam Walters, "The Beginnings," *Mechanical Engineering,* April 1984, pp. 38–46.
32. William S. Pellini, *Principles of Structural Integrity Technology,* Office of Naval Research, Arlington, VA, 1976.
33. Neal FitzSimons, "Engineering Failure—Whose Fault?" *Consulting Engineer,* June 1985, pp. 46–47.
34. Harris R. Greenberg and Joseph J. Cramer, *Risk Assessment and Risk Management For the Chemical Process Industry,* Van Nostrand Reinhold, New York, 1991, pp. 3, 15–29, 224.

Chapter

3

Strategies That Work

Systems-Based Failure Avoidance

Needed: Aids to decision-making

Effective failure avoidance must deal with the entire spectrum of causal elements. It must include human errors and environmental effects, as well as traditional physical causes such as technical faults and malfunctions, inadequate design, and materials-related deficiencies. As said earlier, even apparently simple and inconsequential failures can be complex events, having multiple and interrelated causes. This is where combinations of conditions and events converge in time to precipitate incidents characterized by unattained goals, degraded performance, disappointment, unsatisfactory outcome, and, sometimes, physical damage, economic loss, and human tragedy.

Failures differ in magnitude largely in amount of energy or material that went out of control, size of affected area or number of people involved, and costs of restoring order. There is more public awareness of catastrophic occurrences because of their magnitude and the attention given them by news media. But their causation scenarios are not sufficiently different from those of subcatastrophic incidents.

Failures are unplanned and unexpected occurrences, although high-risk activities have greater propensity for them than do activities of lesser risk. Under these circumstances, there may be more anticipation for problems. Often, however, risks are misjudged. Situations are different from expectations and people's actions and responses are not predictable. Since human beings are not skilled at predicting future events, we do not foresee circumstances linking up with other conditions and events until that critical mix occurs when the failure or accident is, essentially, inevitable.

Our perceptions of these things, before the fact, are clouded and distorted by pressures of the moment, a drive to achieve some goal or meet a deadline, or similar preoccupations. When the failure occurs, the fateful scenario often suddenly becomes clear. Errors, misjudgments, and thoughtless assumptions are now seen in the proper light. It often does not take a mental giant to see instantly how and why things went wrong, whereas, only moments before, there may have been no hint of impending trouble.

This is not always the case. Perceptions of risk and danger, insights into possible scenarios in their embryonic stage, and other avoidance instincts can be developed through understanding, experience, judgment, and common sense. The unfortunate fact is that it takes significantly greater skill to foresee these situations than it does to understand them once they have occurred. Much of the reason for this has to do with complexities of the causation scenario and the fact that, before the event, many causal elements are just not evident. Things are not always what they appear to be, but we usually must accept things at face value.

Hindsight acuity is a useful attribute in failure analysis, in determining what occurred and why, but is of limited value in future avoidance. Of course, once we have correctly diagnosed the causes for some failure, we can apply what was learned to similar equipment, similar products and processes, and similar situations. Frequently, we realize that if things are left to themselves, recurrences are possible; so we make whatever changes are necessary to avoid them. If we have correctly assessed the original failure, we probably have a good chance of successfully avoiding future ones. But, if we missed the point somehow and failed to perceive the true causation network, we may not prevent recurrences at all. This is because we have not identified and eliminated basic causes, only addressed symptoms.

Recall the hydrogen embrittlement of high-strength steel bolts of a previous example in Chap. 2. Even though the bolts were fracturing in service (the symptoms), the cure was not in using higher-strength bolts, which only worsened the situation as their time to failure became even shorter. This was because the underlying cause had not been eliminated. The same was true for the gasoline tank that leaked at a welded seam. Rewelding did not solve recurring leaks. Although the leak site was a weld, its cause lay elsewhere.

Even when we have benefit of hindsight and have had an opportunity to examine failed parts and everything else associated with the failure, we can still miss the mark and not avoid recurrences. But it is far easier to comprehend the failure and to unscramble its tangled causal web after the occurrence than it is to be able to see it coming, to sense the gathering storm, and perceive that trouble is brewing just ahead.

This difficulty imposes a severe limitation upon failure avoidance attempts. If it is difficult to identify correct causes for relatively simple failures, what chance is there to forestall them? And if we are often stymied by simple failures, what of complex ones? The answer to this dilemma does not lie in more sophisticated diagnostic equipment, although it can be useful to have. And it does not lie in theoretical probing into material degradation phenomena.

What is needed is a tool to enable us to comprehend and account for the frequently large number of causation elements of all kinds and their interrelationships more clearly. It should be able to show which elements are most critical and deserving of priority attention. The tool should be versatile enough so that it can be applied to situations before failures, in forestalling them. And it should be applicable to failures that have already occurred, in determining their causation sequence so that effective steps can be taken to avoid recurrences—not only of the same kind of failure, but of similar ones as well. Fortunately, such tools exist.

They are not devices that plug into the wall, but analytical procedures, ways of thinking about things. This methodology, known as *systems analysis* or the *systems approach,* has been under development since the mid-1940s.[1–4] It was developed in response to the recurrent need for improved techniques for objectively analyzing complex problems. Traditional intuitive reasoning is grossly inadequate for the increasingly difficult problems facing industry, business, government, and society in general. There has been greater uncertainty over economic matters, while payoffs of good decisions have improved.

Delineation of the "best" alternative among many options has grown more difficult. Today's important decisions involving commitments of vast resources must address a frighteningly complex interaction of physical, social, political, economic, and technological issues. Many of these issues are critical to a successful outcome of the decision, yet defy quantification, are shrouded in uncertainty, and often have no "best" or optimal solution.

When approached on traditional grounds, difficult or imponderable issues are often assumed away or ignored. As a result, resource projections for achieving desired objectives often vary across broad ranges, and budget overruns become commonplace. Decisions frequently produce negative results, answers to wrong questions, treatment of symptoms instead of causes, or costly attempts to find solutions to problems that can have none.

The systems approach provides a logical framework for inserting all available information, data, experience, judgment, and even intuition, as an aid in making decisions. It helps the decision-maker think about the right things, *all* the right things. The approach cannot tell the user what to do, but it does require him or her to enumerate alternatives and to inquire what is being attempted. In addition, it provides a list of what is needed for a rational decision. It is a schoolmaster of sorts in keeping the decision-maker on the right track.

Early critics of the method called it "common sense made difficult." The method is a common sense approach that extends our innate analytical skills to complex situations involving numerous elements and alternative options—situations where ordinary unassisted intuitive reasoning would be incapable of handling the workload. Systems analysis models, whether mathematical, physical, or graphic, are not much different from mental models that we use in solving everyday problems. The major difference is that systems analysis models are explicit and can, therefore, be manipulated more easily and constructed as a more comprehensive representation of the real world than the subjective models most of us use in solving problems. Not a substitute for common sense and reasoning, it merely enables the user to apply these skills to complex problems in an orderly way, solving a piece of the puzzle at a time.

The systems approach

> ...concentrates on the analysis and design of the *whole,* as distinct from the components or the parts. It insists upon looking at a problem *in its entirety,* taking into account all the facets and all the variables, and relating the social to the technological aspects....the word "systems" connotes the whole, the combination of many parts, a complex grouping of men and machines.[5]

It is a technique ideally suited to analysis of failures and situations having failure potential. It accomplishes one of the principal tasks of effective failure avoidance in that it considers the entire network of causal effects instead of concentrating upon the more tangible, physical, and more familiar technical aspects.

In a way, the techniques really are not new. It has been pointed out that major engineering achievements over the years would not have been feasible without applications of such methods.[6] Examples are telephone and electric power distribution systems, the Panama Canal, the pyramids, and, more recently, space exploration and weapons systems. It is evident that during recent decades engineers have adopted these analytical methods as needs have arisen. Their adoption has been such that the techniques are no longer singled out for special mention as analytical tools, but have gradually become assimilated into the analysis, design, development, operation, and application of new technologies on many fronts.

They are finding increasing use in failure analysis and prevention for major components of critical systems, nuclear power plants, for example. A point often missed, however, is that these techniques can be applied usefully to *any* failure situation—before or after—and even for relatively simple cases. Their methods help ensure that all relevant causal factors are considered and are accorded proper emphasis. This is a major step toward avoiding recurrences.

Broadened scope and outlook—A must

A weldability problem. The need for expanding our scope of attention in analyzing failures is illustrated in a couple of failure examples from the metals industry. Some years ago, an increasing number of weld failures occurred in dissimilar joints between a nickel–copper alloy and alloy steels. This combination is usually welded using nickel filler metal. This is because it is metallurgically compatible with both joint members and its mechanical and physical properties also match, providing the desired joint efficiency.

The failures were characterized by cracks along the fusion line on the nickel–copper alloy side. When the parts were sectioned for examination in attempts to determine the cause, the cracks were often found to extend completely through the thickness of the member (as much as an inch or more in some weldments) and along the entire joint length. This presented serious problems, as many of these joints involved critical naval service and heat exchanger use in marine environments—applications requiring high-integrity welds. Such extensive cracking caused major leaks and loss of structural integrity. Figure 3.1 shows photomicrographs of weld interfaces for both cracked and uncracked welds in the same alloy type using nickel filler metal.

As for many welding problems, the occurrences were sporadic. Not all such welds were afflicted. In fact, many were not, but were sound and satisfactory in all respects. Attempts to correct it focused upon the welding operation. Practically every conceivable variation in welding technique was tried without success. The attempts included changes in welding speed and bead sequence, use of preheat or not, welding power variations, changes in joint geometry, various cleaning methods, joint restraint variations, and many others including use of different welding processes. None of these measures offered promising leads as to the cause or its cure.

Other welds (nondissimilar welds) in the same materials almost always were unaffected. Only those made with nickel filler cracked. Filler metals from different sources were checked, but with the same outcome. Some welding personnel began to believe that the occurrences were due to natural metallurgical phenomena that were characteristic of the binary alloy system. They reasoned that the dissimilar combination was probably always susceptible to cracking if the stress level and degree of heterogeneity of filler and base metal mixture at the interface were "just right." If this was so, a search for a solution to the problem was futile. Indications were that a new approach for making dissimilar welded joints of this alloy combination had to be developed.

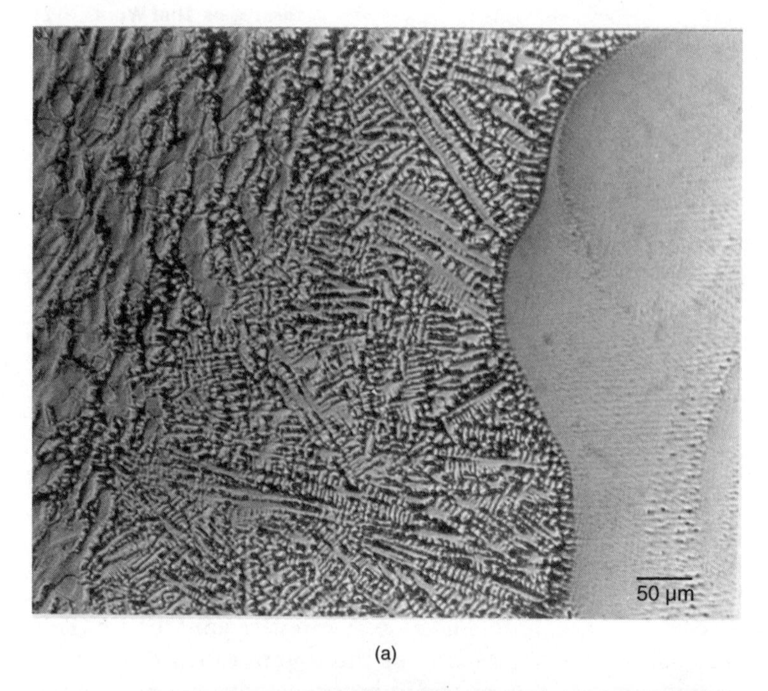

(a)

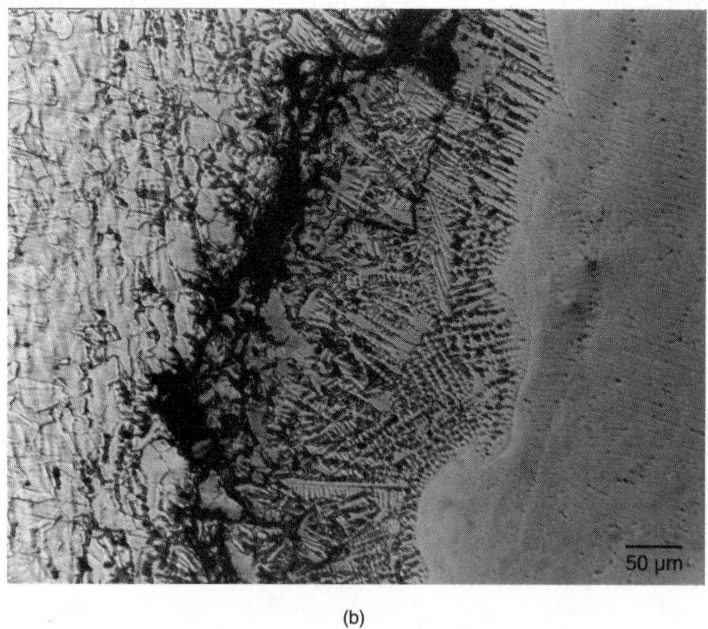

(b)

Figure 3.1 Photomicrographs of transverse cross sections of dissimilar welds in a nickel–copper alloy made using nickel filler [region shown is the interface between the nickel weld (right) and nickel–copper alloy (left)]. (*a*) No fusion line cracking. (*b*) Fusion line cracking.

In time, persistent fabricators devised methods for minimizing the occurrence. Their methods involved weld-overlay buildups (so-called "buttering") along the nickel–copper alloy member, of matching composition weld metal, then remachining the joint configuration, followed by welding as before. The technique was not 100% reliable but did permit some components to be completed. It was a time-consuming approach, as the modified weld procedure required special approvals and certifications, additional inspections, and the extra welding and machining operations. The technique did offer relief but, more importantly, provided investigators with a clue to its cure.

The clue was that since "buttered" edges insulated the nickel–copper alloy base metal from the nickel weld deposit, some aspect of the nickel–copper alloy (plate or tubing) may be contributing in some way. At this point, investigative attention turned to the alloy itself. This was done somewhat reluctantly because all material that had been used for these applications, whether it had cracked during welding or not, had been ensured to be within tight specification limits and had certification records proving full qualification to applicable code requirements.

During this portion of the study, samples of archive material produced years before were located, evaluated, and chemically analyzed for a wide range of elements. Microstructures were examined, and mechanical tests were run. Results were compared to those for more recently produced material, as well as to those for other lots of the same alloy that had been produced during intervening years. Evaluations showed that recently produced lots were prone to crack when welded with nickel filler metal while others were not when welded under identical conditions.

When these results were correlated with trace-element content, it was found that crack-sensitive lots had slightly higher levels of phosphorus, but still well under specification limits. Experimental alloys of controlled compositions, having incremental additions of phosphorus from zero to levels above those found in recently produced material, confirmed a phosphorus threshold below which fusion line cracking would not occur when welded with nickel filler metal. Above that threshold, cracking occurred.

This "answer" seemed technically correct but difficult to understand. Phosphorus had not been a deliberate addition by alloy producers. So, why were phosphorus levels higher in recent lots than in earlier lots, and why did they appear to be increasing?

Subsequent study revealed that alloy producers frequently used selected scrap copper for melting stock. The grades were screened to avoid contamination from solder, lead, and other elements known to be harmful, and all ingots were checked for compliance with applicable specifications before their processing to desired mill forms. It turned out that scrap copper tubing was often part of the remelt charge and this material is deoxidized with small amounts of phosphorus when it is manufactured.

Calculations showed, however, that even if the copper constituent of the nickel–copper alloy had been derived 100% from scrap copper tubing, the phosphorus content would not have been high enough to exceeed the indicated threshold. It was discovered, however, that over a period of time the in-house copper-containing melt stock tended to increase in residual phosphorus. This was due to reuse of off-composition ingots and those discarded because of misruns and remelting of hot-tops (upper portions of ingots containing shrinkage voids) and other such materials that were otherwise reusable. Since phosphorus was not lost during the melting process, it tended to build up within the scrap circuit such that it would, eventually, reach the level that would make the alloy of which it was a constituent susceptible to cracking during dissimilar metal welding.

The point of this example is that the cause of the weld failures lay far from the immediate welding operation itself, which was the principal and obvious focus of the investigation. Much time and effort were spent in examining welds, trying alternative welding techniques, and devising schemes for correcting the problem. Since the material being welded repeatedly met quality specifications and code requirements, it had been dismissed earlier as a suspect.

The need for dissimilar joints contributed to the problem and brought it to light. The higher melting point of the nickel filler weld aggravated the situation in subjecting phosphorus-rich liquated films (along the lower-melting-point nickel–copper alloy interface) to tensile stresses from the solidifying and contracting higher-melting-point nickel welds. Use of weld filler metals having lower melting points would have presented other problems such as intolerance for iron dilution from the ferritic alloy member, inadequate mechanical properties, or corrosion susceptibility.

Here, a specific application required specialized joining procedures and filler metals. Established techniques and procedures had been developed and certified for making sound, crack-free welds for critical service. After extensive testing and field evaluation, the materials and procedures had been approved for the intended service. Nevertheless, occurrences of weld failures cast doubt upon the procedure, application, and materials involved. Unnoticed buildup in the ingot producer's scrap circuit of a residual element commonly found in copper-containing alloys turned out to be the culprit.

If investigators had been content to confine their attention to the apparent cause of the problem—something within the welding operation—a solution might never have been found. Metallurgical experts even had logical explanations relating the occurrence to natural phenomena resulting from inherent heterogeneities of nickel and copper constituents. The problem "system" was thought to be confined to the welding operation and directly related variables.

Yet, the "system" was considerably broader. It not only included welding operations and the requirement to use a dissimilar-metal filler of higher melting point, but also the trace-element buildup within the melt shop's scrap circuit. It probably extended further than that, into the decision to use certain proportions of scrap material in melt charges. Once the cause was identified, the solution was straightforward and effective: establish a new and lower threshold for residual phosphorus in this alloy and monitor it.

An electrical conductivity problem. In a somewhat similar example, a copper producer found that its wirebar production (slated for use as electric motor windings) suddenly would not pass basic acceptance tests. These tests, an industry standard at the time, included evaluation of the behavior of a simple coil spring formed of the material. The spring was wound and annealed under carefully controlled conditions, and cooled. This would soften the copper. The spring was suspended from a fixture and a precision weight attached to its lower end. Extension of the spring and its permanent set under the load were a function of the purity of the copper, as minute residuals would affect its annealing response and show up in resisting spring elongation. The annealing cycle was fixed by conditions of the wire enameling process and was not subject to modification. For motor windings, springback response and electrical conductivity were important, accounting for these acceptance standards.

The source of the apparent contamination was a mystery. All conceivable sources were believed to have been checked. Once again, archive samples retained by the refiner's laboratory provided clues. Analyses showed that the contamination had not occurred spontaneously, but had been gradual, beginning at some time several years earlier. There had been a slow buildup of various trace elements within the wirebar copper that had reached a sufficient level to be detected in this sensitive test. What was the reason for the buildup?

A comprehensive survey was conducted for the entire plant, including its history and operations over the period of trace-element buildup and for a year or two before it began. It was discovered that the buildup began just after new environmental regulations imposed restrictions upon smelter emissions. To comply, more efficient scrubbers, dust collectors, and precipitators had been installed in the smelter stacks; this was just before the time when the contaminant buildup began. Because of sizable quantities of accumulated metal-containing dusts that had been recovered from the new collector systems, and in the spirit of resource conservation, it was decided to reuse these quantities of stack effluents as smelter feedstock, to be subsequently refined.

Once more, the trace elements did not go away, but were simply recycled over and over and, with each recycling, additional amounts from raw materials were added to the total, resulting in a gradual buildup in the refined product. Here we have an installation of a new efficient dust collection system creating a problem for a metal refinery located miles away. It was not until "system" boundaries were extended well beyond the confines of the immediate problem that the cause was identified.

These examples, only two of many, serve to reemphasize the complexity of the causation network and the need for a broad outlook when evaluating failure propensity and causes. The system being evaluated must include all factors that can influence success or failure of the plant, product, equipment, or whatever is being considered or studied. Where should one locate the boundary and draw the line? Obviously, everything is in some way connected to everything else. Consequently, some restraint is necessary or the "system" will be of tremendous size and therefore unworkable. Judgment is required in setting system boundaries. They need not be cast in concrete but can be moved as the study progresses to suit the subject under consideration. The boundaries usually lie well beyond the area of the immediate problem.

Herbert Spencer, the English philosopher, said, "When a man's knowledge is not in order, the more of it he has, the greater will be his confusion." It will become evident, if it is not already, that the systems approach is primarily a mental discipline, although it is much more. When analyzing failures or potential failures, particularly for complex engineered systems, it is easy to become confused and overwhelmed. Systems techniques help keep knowledge and facts in order and thereby lessen confusion.

The decision-making process

Each problem and situation has its own set of constraints, requirements, and approach. Steps for resolving a design problem will differ from those for determining optimal operational parameters for a chemical process, for allocating capital budgets for a manufacturing plant, or for deploying missiles for maximum defensive effectiveness. Our interest is primarily concerned with avoiding failures. Therefore, an initial decision is to identify components, processes, subsystems, or actions that are critical to their continued safe operation and to locate potential trouble spots.

Implementation of avoidance strategies discussed in subsequent sections requires allocation of funds and other resources, and it is important that these resources are spent on the most critical items. Therefore, these choices for resource allocation should not be made lightly, intuitively, or subjectively. This selection should reflect accurately the potential for damage, injury, and economic loss that could result from a failure. This is not always immediately evident and may require a thorough analysis.[7]

Figure 3.2 outlines basic steps of a decision model for such a situation. The steps outlined are shown in a logical sequence, although they need not be rigidly chronological. They are the same fundamental steps we use every day in analyzing situations without being consciously aware of their content, progression, or interrelationship.

Input can come from a number of sources, as indicated. Methods for developing this input, including information on system hazards and faults and their relative criticality, are described in the following section entitled *Predictive Modeling and Analysis.* If relevant failures have occurred, results of their analysis would also be used as input for the decision model. The feedback loop from the sensitivity analysis step to previous stages reflects the iterative nature of most decision-making

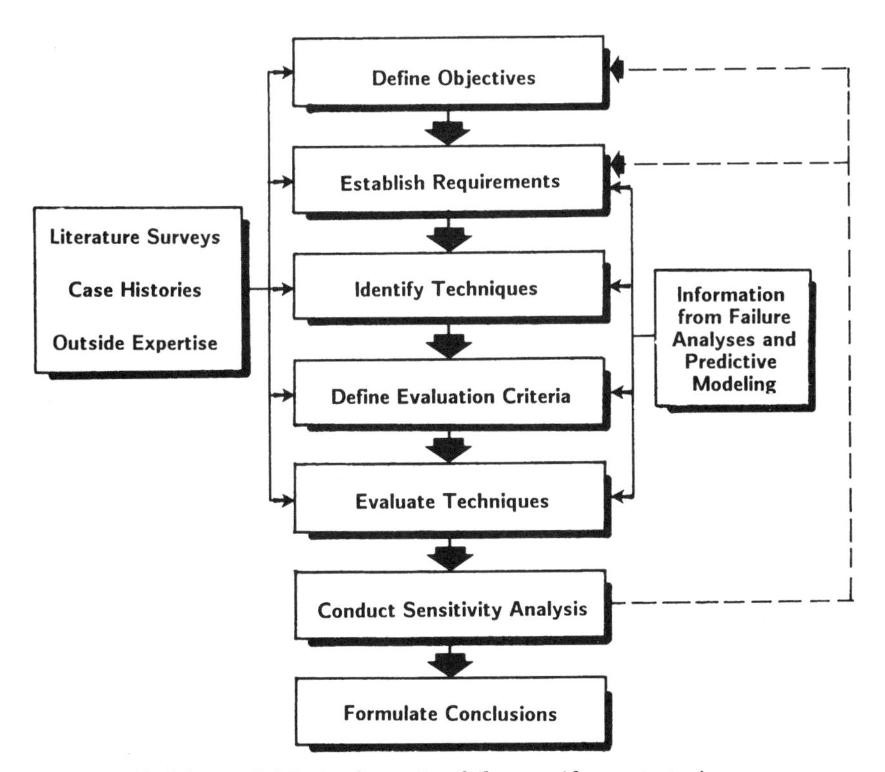

Figure 3.2 Decision model for implementing failure avoidance strategies.

processes when results of initial trials may not provide satisfactory answers, prompting further analysis.

The purpose of the decision model, as presented in Fig. 3.2, is to determine how best to apply available techniques and avoidance strategies to avoid failures of some system, assembly, component, part, or product. This is the objective. To do this, it is essential to identify the most susceptible component or the component or subsystem that is most critical to maintaining safe uninterrupted operation.

The next step establishes requirements to fulfill the objectives and will depend upon the situation and the kind of event that is to be avoided. It may mean that failure of a check valve *must* be communicated to the system operator or that its failure should somehow be automatically compensated for by the system. It may mean that a leak of some reagent *must not* be released to the atmosphere or that some component should never become loaded beyond some set value. These are the requirements for ensuring continued safe and uninterrupted operation. They can take many forms and approaches depending upon the system involved, its hazards, and performance expectations. Input from failure analyses, hazards analysis, and fault-tree and failure modes and effects analyses (discussed in a following section) are essential in establishing requirements.

Alternative techniques for achieving these requirements and objectives also must be identified. These would be avoidance strategies. For existing and operating equipment, they may be some kind of life-cycle management procedure, component-monitoring technique, or a combination of them. They may be other methods, including redesign, materials substitutions, improved communications, or more training. For equipment or systems under development or in the design stage, the strategies would be different.

Avoidance methods are evaluated against selected criteria for judging effectiveness in fulfilling requirements and achieving objectives. The criteria might include such factors as component criticality, risks and consequences of a failure, costs, safety characteristics, maintainability, reliability, accessibility, control capability, technological maturity, and ergonomics (the person–machine interface). They may also include interfaces with other components or assemblies and environmental effects. Tabular formats are used in rating candidates against selected criteria.

With requirements established for achieving the failure avoidance objective, techniques identified for meeting these requirements, along with criteria for their evaluation, it should be possible to assess their relative performance. This step can narrow the number of avoidance methods. And, it provides an opportunity for assessing the analytical procedure so far. It may be found that the evaluation criteria or their assigned performance weightings were incorrect. Perhaps other criteria should have been included, or new techniques may have come to light and should be evaluated. If so, previous steps can be repeated with this improved and updated input. The analytical procedure should continue in this manner until the input and results fairly represent the actual situation, at least to the extent that can be determined.

At this stage, it is advantageous to evaluate possible effects upon the performance ratings of changes in assumed characteristics. That is, subtle changes from assumed properties or capabilities of one or more alternative techniques may have unanticipated major effects upon the outcome. This step, known as *sensitivity analysis,* considers and evaluates deviations from assumed characteristics of alternative failure avoidance techniques.

This information is used as input for reevaluation. If it is found that the results are substantially affected by minor differences in assumptions, more accurate information may be needed to ensure that the performance ratings for competing approaches are valid. Or, if better information is not available, this should at least be factored into the selection step.

The following sections describe techniques for identifying hazards and faults in a system, equipment, or product and for evaluating effects of causal elements upon failure occurrences. This information is needed for implementing failure avoidance strategies for both operating equipment and systems under development.

Predictive Modeling and Analysis

Purpose and function of predictive models

Not long ago, designers had to rely solely upon their engineering skills, judgment, intuition, and personal experience in designing reliable, safe products and systems. Use of generous safety factors and other conservative practices were the principal tools for avoiding failures. This often resulted in overdesign and other inefficiencies. Failure reports offered limited guidance whenever details were available. For most failures, however, accurate information was accessible strictly to in-house staff. It seldom found its way to others who might have benefited. Even today, this situation still exists. Also, because of specialization, accounts of failures and their causes in one industry may not be relevant to those outside it. This is so, even as failure analysis information is becoming increasingly available through seminars and published case histories.

The models and techniques described in the following pages fill the need for reliable guidelines for avoiding failures, as they provide a means for independent assessment of hazards and faults inherent within a specific plant, piece of equipment, or individual component. And they are adaptable to virtually any industry, product type, or form.

In this chapter, we are dealing primarily with operating equipment and processes, products in use, and in-service systems: some may be just starting up, others may be operating well beyond their intended lifespan, and everything in between. The analytical methods discussed are certainly applicable to conceptual and design stages as well, and should find use there and throughout the design–development–production cycle. These aspects are addressed in Chap. 4. Modeling procedures may not differ appreciably from operating equipment to equipment under design, but application of their findings will differ, along with avoidance strategies.

Hazards analysis

Each plant, facility, operation, or installation has its own set of hazards and potential problems. This is where failure avoidance efforts must start—with identification of these hazards.[8–12] To be consistent with our broadened definition of *failure* in Chap. 1, a hazard would be any condition that had a potential for causing one or more of these unwanted situations, incidents, or events to occur. Some will be more serious than others in terms of probable damage, economic loss, and human injury. Each company or location must make its own assessment as each operation and installation is unique. Hazards definition demands careful consideration. This is a judgment matter and no amount of analysis or modeling can compensate for failure to identify inherent hazards in an operation, process, equipment, or system.

Hazards assessments require detailed information about everything going on inside the organization or within its system boundaries. Although the primary focus may be some particular operation, interactions among operations and processes require a broad perspective. Hazards screening must use up-to-date floorplans and equipment layouts; production, process, and material flowcharts; in-process and raw material inventory lists; personnel information, and much more.

The first step is to recognize and identify hazards and potential trouble spots. A tabular format of some kind is needed for documenting this information. It may be preferable to model it after a production line sequence or other flowchart to verify that each stage and operation has been considered.[13,14] Here, again, checklists will ensure that all important details are included and considered. Space should be provided in the format for notes on measures for diminishing or eliminating identified hazards. Hazard indices, or numerical ratings denoting severity, may be useful in establishing relative risks and priorities for devising failure avoidance strategies.

Various hazards analysis approaches are used. One suggested method[15] identifies the following:

1. Hazardous elements (e.g., pressure vessels, toxic material containment, combustibles storage)
2. "Triggering" events that can transform hazardous elements into hazardous conditions
3. Corrective measures (eliminate the condition or guard against it)

Another[16] describes seven steps in hazards analysis. They include reviews of the following:

1. Past experiences
2. Performance requirements
3. Operating environments
4. Methods for hazards elimination or control
5. Methods for minimizing damage
6. Implementation techniques and procedures
7. Implementation responsibility

A matrix chart for conducting preliminary hazards analysis might have the basic format shown below. Under the headings would be brief notations for each identified hazard:

Condition	Cause	Consequences	Category	Correction
Gasoline tank leak on boat	Corrosion Fatigue Bad weld Loose fitting Defective line External damage	Possible fire and explosion; injuries (perhaps fatal) and equipment loss	I - II	Check for presence of corrosion potential (dissimilar-metal couples) Inspect and leak-test when installed; periodically remove, drain and clean tank, reinspect and retest

Hazards analysis notations may be brief, as in the preceding example, or detailed, depending upon the situation and resources available for the job. (See the failure modes and effects analysis example on p. 85 for an explanation of category notations.) It is preferable to start with basic simplified formats and a qualitative approach before progressing to more elaborate matrices and attempts to quantify results. Hazards analysis is not an end in itself, but an important initial step toward gaining a grasp of the magnitude and extent of failure potential within a given facility. But, whatever the format used and however extensive its detail, it should be well-documented, together with references to supporting information, data, and input sources.

Fault-tree analysis

Hazards analysis identifies hazards and potential failure sites but does not usually furnish the kind of detail needed to pinpoint specific components and their flaws, although it narrows the field. Effective failure avoidance strategies require more precision, and this may be obtained through various methodologies that have been developed.

One method, fault-tree analysis (FTA), has already been mentioned briefly in Chap. 2. FTA is a deductive logic model that depicts, in graphic format, conditions and combinations of conditions that can produce the "fault" or failure under consideration. It models system conditions that could lead to trouble and, in doing so, reveals interactions among hardware, people, and environment, as well as interrelationships among causal elements.[17–26]

As usually constructed, the failure, fault, or other unwanted event is shown at the apex of the "tree" with branched links below, through logic gates, to causal elements or conditions. FTA is true analysis, as it starts with the event and dissects it into its component parts. The value of this technique, as for all systems techniques, rests heavily upon the quality and validity of its input plus user skill, experience, and judgment. Of course, the model cannot provide better information than the user possesses and incorporates into its construction, although it can arrange user information and input such that the user's vision and perspective are improved and his or her insights are clearer and more accurate.

Following up on the simplified fault-tree example of the preceding chapter (Fig. 2.2) is a more expanded fault tree (see Fig. 3.3) for the same gasoline tank example. This version considers additional failure modes and causes, although others might have been included. (Figure 3.4 describes the graphic symbols.) From this illustrative example, it should be evident that such a representation can be useful in analyzing failure propensity for any system, subsystem, assembly, or piece of equipment.

An engineered system may have any number of faults and potential failure modes and each would have its own fault tree. Despite the apparent simplicity of the illustrated examples, fault trees can be extremely complex, particularly when they are used in developing mathematical probabilities (using set theory and Boolean equations). Computer software developed for these purposes is virtually indis-

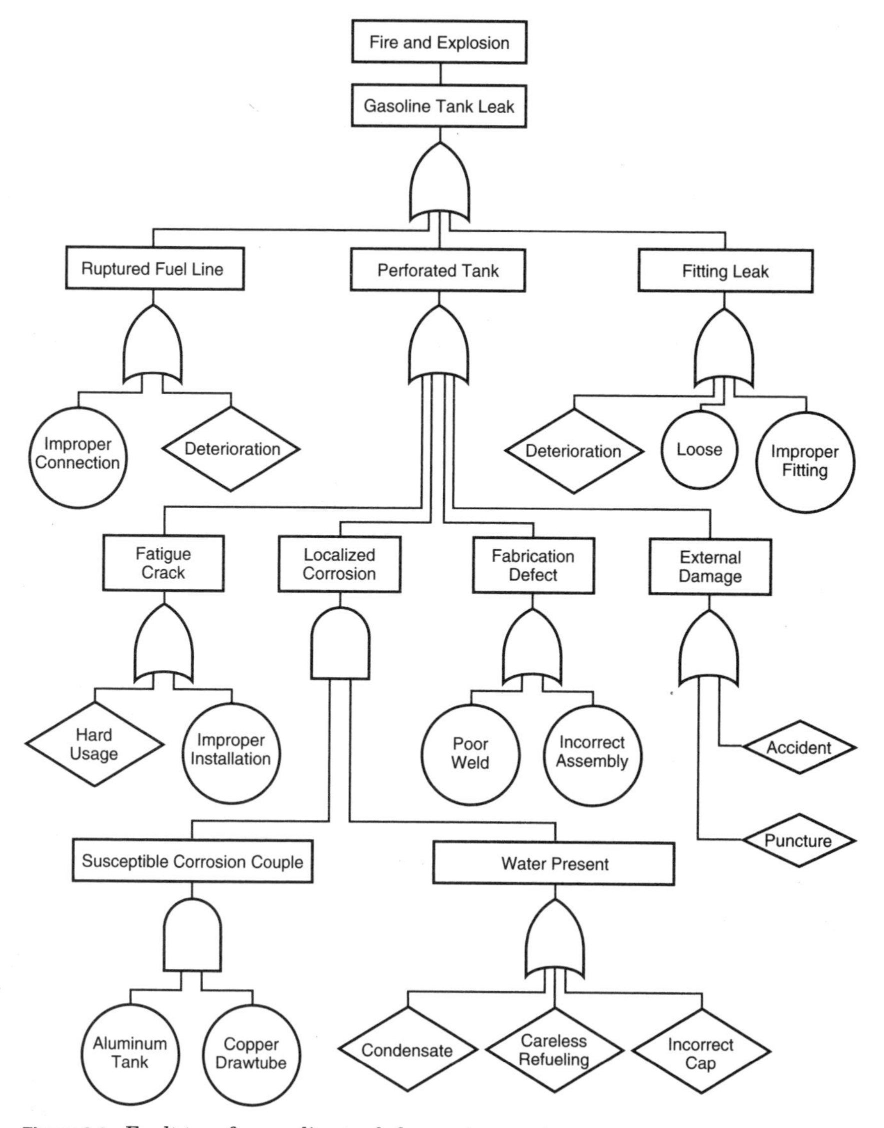

Figure 3.3 Fault tree for gasoline tank for marine service.

pensable in facilitating analysis of extensive fault trees.[27] There are disadvantages in attempting to simplify fault trees for inherently complex situations, as important details can be missed, but it is advisable to learn these techniques using simplified and familiar examples.

Fault-tree construction and analysis for major engineering projects and systems are best left to specialists with experience in this activity, proficient in its mathematics and knowledgeable in the availability and

EVENT Symbols

Event

Identified fault or event, produced by combination of other more basic causes; a "gate" input or output.

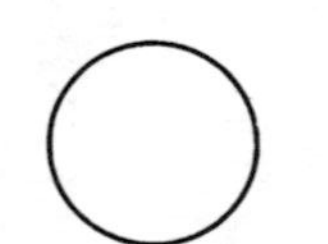

Primary Failure

Basic events, malfunctions, or causes, with available data from test results or failure analysis information.

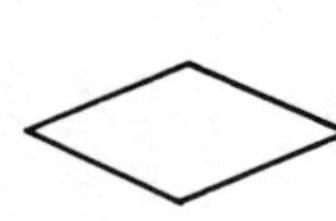

Undeveloped Event

Events not pursued to determine more basic faults, used mostly for information purposes; limit of fault tree resolution.

Normal Event

A system characteristic; event presumed to always occur.

GATE Symbols

AND Gate

Output event exists or occurs only when all input events exist or occur simultaneously.

OR Gate

Output event can exist or occur if any one or more of input events exist or occur.

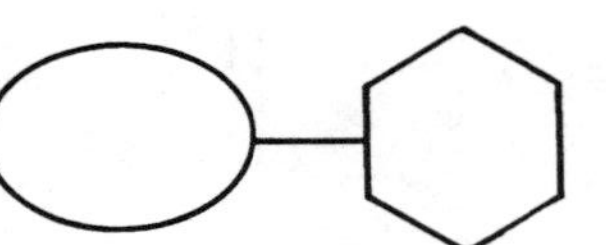

Inhibit (or Conditional) Gate

Output is conditional, depends upon occurrence of restricting or qualifying event; controlling variable is described in adjoining symbol.

Figure 3.4 Fault-tree symbols.

application of computerized techniques. Nevertheless, all concerned with failure avoidance today should become familiar with this tool. This is best accomplished through actual hands-on construction and analysis of simplified models and perusal of the cited references.

Failure modes and effects analysis

Another analytical technique, often used in conjunction with FTA, is failure modes and effects analysis (FMEA).[28–32] Its logic is the reverse of FTA in that the FMEA format, really a synthesis, starts with the most basic components and works its way down into more complex assemblies and subsystems. Its content is usually depicted in a spreadsheet resembling an accounting or inventory form. It is hardware-oriented, listing each part, and evaluates how each might contribute to a failure, malfunction, or other undesired condition or event.[33,34] FMEA lists minor components first. However, in practice, the analyst usually begins with entire systems, then subsystems, followed by more basic components and individual parts, by serial number. In the FMEA format, columnar headings might be as shown in the example below:

Part	Function	Failure						Minimization Measures
		Mode	Cause	Result	Consequences	Severity[(a)]*	Probability[(b)]*	
Fuel tank	Contains gasoline	Fuel leak	Corrosion Fatigue Loose fitting Defective line External damage Bad weld	Fire and explosion	Injuries (possibly fatal), equipment loss	I - II	C	Check for corrosion potential (dissimilar-metal couples) Inspect and leak-test when installed; periodically remove, drain and clean tank, reinspect and retest Check fittings for tightness and lines for deterioration; periodically replace lines

(a) *Severity Categories*: I - Catastrophic, II - Critical, III - Marginal, IV - Negligible

(b) *Probability Codes*: A - Frequent, B - Reasonably Probable, C - Occasional, D - Remote, E - Extremely Improbable, F - Impossible

[*Adapted from MIL-STD-882A, *System Safety Program Requirements* (28 June 1977) 11-12.]

FMEA input is derived from parts lists, assembly charts, equipment diagrams, failure analyses and other records, and analysts' experience and knowledge. FMEA formats can be tailored to suit specific situations and should include as much detail as is necessary for comprehensive evaluation of the selected system or component. Entries might include, in addition to those shown in the example, failure symptoms, service life expectations, failure history, and reliability projections. The example shows only a single component, the tank assembly, but an actual FMEA would include entries representing a more detailed breakdown including fittings, caps, vents, fuel lines, gauges and sending units, gaskets, hold-down fixtures and assemblies, and any other parts or hardware.

FMEA is a time-consuming and painstaking task; however, its benefits can be significant. Frequently, it identifies seemingly incidental items as being critical to successful and continued performance of major components. More attention to the quality and integrity of these apparently insignificant items can often prove highly cost-effective in avoiding failures. Such findings can make the expense of FMEA and other analytical techniques a worthwhile investment.

Both FTA and FMEA reveal how even minor defects can contribute to potential failures of major consequence, and interrelationships are often clearly seen. But FTA offers a perspective of the system that may not be obvious in the FMEA format; for this reason, FTA is usually conducted first.

Because nothing remains static but is in a continual state of flux, hazards analysis, FTA, and FMEA can never be considered completed. Raw materials and grades change, substitutions are made, manufacturing methods and processes are modified, throughputs and product mixes change with customer demands, old product lines are phased out and replaced with new, company policies change, personnel come and go, and government regulations can force adjustments in all activities. Therefore, it is a continuing challenge to update these analyses. Results from ones conducted last year may not be relevant this year. Assignment of this responsibility is an important decision and demands full attention of qualified people.

Much of the value of these procedures lies in clearly documenting all steps and their rationale, including all assumptions made and used during the analysis. Such a record can prove extremely valuable later on, as in demonstrating due care in defense of a lawsuit, for example, as well as in providing input for other analyses.

The report describing the analytical process provides rationale for making the final choice, and this information is needed to support resource allocations for implementing study recommendations. Reviewers may disagree with conclusions or object to recommendations; however, a major benefit of this kind of analysis is that the entire decision-making process is open to reviewers. Critics are forced to identify which assumptions or study input leading to the conclusions or objectionable recommendations are questioned. This analytical methodology tends to discourage generalized criticism and, therefore, facilitates the process, its endorsement, and implementation of recommendations.

Principal Damage Mechanisms and Sites

Materials degradation

Plants and equipment age, their performance deteriorates, and they become susceptible to failure for many reasons. Materials degradation, however, is the dominant reason. Materials give objects substance, sustain loads, transmit forces, contain fluids under pressure or vacuum, serve as barriers and containment boundaries, and provide protection. When these degrade in some way from mechanical, thermal, or environmental effects, or combinations of them, or when they are accidentally damaged, they are no longer dependable. If they are allowed to continue in that condition, they can fail or cause something else to fail.

Accordingly, materials integrity is the focal point of life-cycle management or life-extension technology. Of paramount importance, then, is an understanding of how given materials respond under actual operating conditions and to degrading influences—whether they are time, temperature, stress and strain, environment, accidental damage, or all combinations.

Our emphasis in the following discussions will be upon metals and alloys. They are the dominant materials of construction for practically all failure-susceptible components of mechanical systems and they have been the subject of virtually all life-extension activity to date in the major fields. Nonmetallic materials are finding increasing use, however, and have their own failure modes and susceptibilities.[35] Many of the strategies that are being applied to metals are also applicable to nonmetals and composites.

Table 3.1 lists predominant failure or degradation mechanisms for metal and alloy components. It is not an exhaustive list, and failures are often the result of combinations of these individually listed mechanisms or processes. Not all metals and alloys are susceptible to all types. From the list, there appear to be many failure mechanisms, but the material–component combinations where these types of failures can occur are much more numerous, and are practically limitless.

Such an extensive list of degradation mechanisms would convince any skeptic that engineers are working against a stacked deck, where Mother Nature has other plans. These degradation mechanisms and others merely reflect the fact that nature is continually and relentlessly striving to undo our attempts to restore order to the world's disordered system of things (as viewed from the viewpoint of permanence of refined materials). Entropy eventually wins. Refined metals and finely tuned alloy systems that have cost us our resources will one day

TABLE 3.1 Failure Mechanisms for Metals and Alloys

Abrasive wear	Hot corrosion
Adhesive wear	Hydrogen embrittlement
Atmospheric and general corrosion	Intergranular corrosion
Carburization	Lamellar tearing
Cavitation	Liquation cracking
Corrosion-erosion	Liquid metal embrittlement
Corrosion-fatigue	Low-temperature embrittlement
Creep and creep-rupture	Mechanical overload
Crevice corrosion	Oxidation
Delamination	Pitting corrosion
Distortion and deformation	Selective leaching and dissolution
Erosion	Sensitization
Erosion-corrosion	Strain-age embrittlement
Fatigue	Stress-corrosion
Fretting	Stress-rupture
Frictional wear	Surface fatigue
Galvanic and dissimilar metal corrosion	Thermal fatigue

revert back to their original forms that they occupied from the eons of antiquity.

All we are seeing in this list of degradation mechanisms are a few of the approaches in nature's kit for accomplishing this. Our task, in ensuring that our intentions are not thwarted by the natural processes in bringing about a premature reversion to disorder (or, what we term "failure" in our engineered products), is to learn how to supersede or forestall these processes and events. This means that we must develop insights into—and gain understanding of—the processes and mechanisms that can so severely degrade our fine engineering efforts. This is the only approach that will work, and this is the goal of this book.

Unfortunately, failure avoidance strategies and monitoring techniques applicable to operating systems have not been developed for the entire catalog of failure mechanism–material combinations, and perhaps never will. Nevertheless, existing techniques are capable of detecting the presence and degree of virtually all of these conditions in operating equipment, that is, provided that there are sufficient motivation and need for doing so, as all of the listed degradation mechanisms are well recognized and their causes well understood.

What this says is that knowledge and technology probably exist for devising strategies and monitoring techniques for avoiding failures by any of these mechanisms whenever the need for doing so is great enough. A review of developments already accomplished offers encouragement and guidance for those developments not yet achieved or, perhaps, not yet even attempted.

But many of the listed failure mechanisms are unlikely to lead to *catastrophic* failure. This does not say that they do not have the potential for doing so or that, in isolated instances, have done so. For many of them, degradation is gradual and there is ample warning that something is wrong. This prompts us to check things out and avert serious damage. However, a good number of failure mechanisms involve cracking, and this is the area of major concern.

Deterioration mechanisms involving cracking are insidious. Cracks can begin subtly and sometimes incubate over long periods of time, often progressing relentlessly deep within a component. During their stage of slow growth—usually with no observable plastic deformation—there is no warning of their presence. All the while, the load-carrying capacity of the cracked member is slowly diminishing as the effective unit stress in the region is increasing and further accelerating crack progression. When the crack reaches a "critical" size, it propagates rapidly, usually at sonic velocity, until the part no longer can sustain the load and finally fails by mechanical overload. These latter stages can occur virtually instantaneously. This is not always the scenario but it frequently is. Sometimes, this crack development–propagation–failure scenario involves several degradation mechanisms. One may initiate at a favorable site and set the stage for others to take over, with the final incident occurring through yet another mode.

Since cracks within components are so insidious and play such a dominant role in catastrophic failures, most life-extension techniques are aimed squarely at crack detection and monitoring of conditions conducive to crack initiation and growth. Successful crack detection and control in any hardware system can constitute a major step in forestalling failure.

In view of the importance of preventing cracks and control of growth of flaws and defects, the following sections will discuss in some detail a few of the more common types of crack-involved failure mechanisms. Since welds are a favored site for failure and failure initiation, some of the reasons for this will also be discussed. All engineers responsible in any way for the design, operation, or use of mechanical equipment, or engaged in related activities, will be significantly more likely to avoid failures of these common types if they are well informed in these matters.

Metal fatigue is clearly the most frequent failure mechanism, and it is also one of the most difficult to deal with. Despite its predominance, it is described after brittle fracture, stress-corrosion, and hydrogen embrittlement as fatigue issues and methods of control may be more meaningful then.

Brittle fracture

Unstable crack propagation. Although not listed specifically as a failure mechanism, brittle fracture is the failure mode for many of the listed mechanisms and therefore deserves primary attention. The subject is so vast that even a cursory treatment is beyond the scope of this book. The intention here is simply to present basic notions to assist engineers of all disciplines in understanding brittle fracture, how and why it occurs, and approaches for avoiding it.

Brittleness is a relative term for a material characteristic. It refers to fractures that occur with little or no accompanying plastic flow or ductility. Tensile strength is not a useful index for assessing it, as is evident upon comparing levels of energy required to fracture glass and for fracturing aluminum of comparable tensile strengths. Although we often associate brittle fractures in metals and alloys with inherently brittle materials, like glass, pottery, and cast iron, or with low operating temperatures or impact loading, these are not necessarily required.

The principal feature associated with brittle fracture is instability, where an existing crack or flaw suddenly and rapidly propagates under an applied stress that may be considerably lower than the "handbook" yield stress for the material. The initiating flaw may be a fatigue crack, a stress-corrosion crack, weld crack, or interconnecting internal or external defects such as inclusions, voids, folds, or delaminations.

Catastrophic structural failures have been traced to micromechanical events involving minute volumes of metal. Cracks of millimeter dimensions have led to fractures of ships, bridges, and other large structures. These fractures are triggered by the inadequate ductility of small aggregates of metal grains at crack tips.[36] Although it is not possible to determine susceptibilities and probable crack propagation behavior through examinations of microregions at the root or tip of cracks, testing methods have been devised that provide this information. From this, material "properties," reflecting response in these tests, are developed. This and other related issues will be described in the section under fracture mechanics. However, a brief description of events leading up to the development of these techniques will provide background for an appreciation of their contribution to the evolution of structural reliability concepts.

By traditional design practice, structural members are designed on the basis of loads to be carried. Their dimensions are calculated from published mechanical properties of structural materials and from the anticipated loads. Usually, the design stress is chosen to be some fraction of the material's yield strength to provide a safety margin for overloads and other unknowns. If, during subsequent use, the component fails with accompanying signs of plastic deformation or bending, it is apparent that it failed from overload. But what if the failure occurs with no appreciable deformation, and at a stress considerably lower than the component was designed for? How could such a failure be attributed to overload? To what would it be attributed? Such failures are not uncommon.

A common explanation of past decades, and still heard from time to time, was that the metal in the failure region "crystallized," a conclusion based upon the characteristic crystalline appearance of such fractured surfaces. From limited knowledge of metal fracture, this was often the only available explanation, notwithstanding the fact that virtually all metals and alloys in their solid state are crystalline (except for very specialized amorphous forms developed in recent years). The explanation was invalid, offering absolutely no guidance in avoiding recurrences.

In response to the obvious inadequacy of such explanations and the increasing number of brittle fractures associated with welded tankers and merchant vessels (Liberty Ships) during World War II, extensive studies were conducted in an attempt to learn their cause and to devise avoidance strategies. Before the 1940s, brittle fractures had not attracted so much interest, as most large structures such as tanks, ships, and bridges had been riveted or bolted together. In such construction, brittle fracture in one member was generally an isolated event of subcatastrophic significance.

A stunning exception, however, is coming to light even as this is being written (late summer of 1993). Maritime experts studying photographs of the wreckage of the *Titanic*, the British luxury ocean liner that sank in the North Atlantic Ocean in 1912 on its maiden voyage with a loss of some 1,500 lives, are suggesting that brittle fracture played a major role in the disaster. It is speculated from the appearance of riveted hull fragments still lying at the ocean floor that, because of inferior fracture properties and low temperature, the steel was "glass-brittle" and simply shattered under the impact of striking an iceberg. This had been the suspected cause, although any defect exceeding critical size could trigger propagating brittle fractures.

The investigators suggest that a better grade of steel (i.e., of lower nil-ductility temperature) might not have fractured so catastrophically, allowing more time for the damaged vessel to remain afloat and for rescue vessels to arrive before it sank. If the suspected scenario is correct, this catastrophe, like many others, demonstrates the often-disastrous consequences of extending technology (construction of an ocean liner of unprecedented size) beyond the knowledge and competence (with respect to materials behavior) of the builders and users. Recovery of plate sections and their evaluation by today's standards should prove to be an informative exercise.

Increased use of welding in the 1940s produced a new kind of interconnected monolithic structure having metallurgical continuity, and a new threat to structural integrity from brittle fracture. Here, a crack originating in one isolated spot could propagate catastrophically throughout the entire body. This prompted investigators to realize that traditional mechanical property criteria (e.g., ensuring that materials had tensile ductility) were inadequate for assuring structural reliability for steels.

Examinations of brittle failures in welded ships provided significant clues in identifying conditions for initiating, propagating, and arresting fractures. For example, fracture initiation sites often contained minute defects that appeared to have triggered the occurrence. Also, it was observed that various conditions, such as crack entry into regions of low tensile stress or into plates of different composition, tended to arrest the propagating fracture. These observations led to laboratory assessments of impact toughness of various representative plate samples using Charpy V-notch (CVN) impact tests, and their results correlated with observed fracture initiation, propagation, and arrest behavior in affected plates.

Ductile–brittle transition. This correlation prompted development of various fracture tests using crack-starter techniques in attempts to simulate configurations of actual propagating cracks better. Deformation tests conducted over a range of descending temperatures revealed that steels resisted fracture initiation until some low temperature was reached, at which point fracture occurred with no observable ductility—the *nil-ductility temperature* (NDT). These ductile–brittle transition temperature types of tests provide good correlation with fracture behavior in actual situations and offer a basis for selecting, specifying, and evaluating steels for applications requiring resistance to crack propagation.[37]

Several tests also were developed to evaluate catastrophic propagation of cracks in the presence of sharp stress-concentrating notches. Such configurations can make structural materials susceptible to brittle fracture. The tests, carried out on small laboratory samples, were intended to simulate actual fracture site configurations and provide guidance in predicting service performance. These tests are still widely used for these purposes.

In addition to CVN impact and NDT drop-weight tests, there are dynamic-tear (DT), wide-plate, drop-weight tear (DWTT), and notched tensile tests.[38–43] These tests are straightforward, relatively simple, and easily carried out; however, they have limitations. For example, some metals and alloys do not exhibit a ductile–brittle transition. Also, for some applications, it is important to have information on effects of loading rate and quantitative data on crack-growth rates under various loading conditions as in assessing component life under fatigue or stress-corrosion conditions. For these reasons, and others, fracture mechanics has become the basic tool for assessing fracture propagation characteristics.

Fracture mechanics. Analytical procedures for predicting structural behavior of cracked or fractured materials under stress are known as *fracture mechanics.*[44,45] The discipline is not concerned with initial flaws or preventing them, but presumes their existence and evaluates their effects upon the integrity of the component or structure and its risk of catastrophic failure. Fracture mechanics deals with stress analysis of the region around a crack tip (i.e., gross stresses assuming a sharp-edged crack) and it quantitatively ties together stress level, flaw size, and material toughness. These techniques are not applied to inherently brittle materials, but principally to metals and alloys that are ordinarily ductile or at least behave that way until a critical-sized flaw develops.

These methods are the principal means for evaluating susceptibility of materials to brittle fracture and their fracture propagation characteristics; therefore, their protocol and application will be described in some detail. The primary factors controlling susceptibility to brittle fracture are material toughness, crack size, and stress level. Fracture mechanics offers quantitative information on their interrelationships, and tests have been devised to deduce the micromechanical behavior at the crack tip that influences a material's toughness.

In the situation of usual interest, where the crack plane is oriented normal to the tensile loading direction (mode I, where the fractured surfaces separate like pages in a book), a stress-intensity factor, designated K, is used in the linear-elastic analysis of the stress field. That is, K characterizes the resistance of a material to fracture in the presence of a sharp crack under severe tensile constraint. It is a function of crack geometry and the nominal stress acting in the crack region. Crack progression is related to the size or volume of the plastic zone at the crack tip; the larger it is, the tougher the material.

The critical K level that can be imposed upon the crack tip for mechanical constraint of maximum triaxiality (i.e., plane-strain) is defined as K_{Ic} (I denoting the crack-opening mode, and c denoting critical). Values for K_I are given by the equation

$$K_I = \lim_{r \to 0} [\sigma_y (2\pi r)^{1/2}]$$

where r = a distance directly forward from the crack tip to a location where the significant stress is calculated and σ_y is the normal stress. The critical plastic zone size is a function of $(K_{Ic}/\text{yield strength})^2$, and this ratio defines the material's fracture toughness.

Fracture mechanics deals with three basic fracture states: (a) plane-strain, (b) elastic-plastic, and (c) plastic. Plane-strain applies to materials characterized by linear-elastic behavior. That is, fracture that occurs under elastic stresses with little or no shear-lip development (i.e., essentially brittle). The elastic-plastic state exhibits a mixed-mode behavior and is typical of many alloys of engineering interest. Most structures are constructed of materials that exhibit some level of this behavior at service temperatures and loading rates. Plastic fracture occurs in ductile materials and requires loads exceeding the yield strength; consequently, these kinds of materials are seldom, if ever, involved in brittle fractures.

Tests used to develop fracture mechanics information (i.e., numerical values for fracture toughness) must be conducted under rigidly prescribed conditions appropriate to the fracture state (e.g., plane-strain or elastic-plastic) if the values are to be structurally significant. To ensure this, standardized testing procedures should be rigorously followed.[46–50] Where test conditions do not comply with the standards, K_{Ic} is "apparent" and has only limited significance. K_{Ic} is most appropriate in thick material of high constraint, and large test specimens are usually required for obtaining valid data.

The K_{Ic} parameter is used for plane-strain (linear-elastic) materials; whereas J_{Ic} (sometimes referred to as the J integral) and other techniques recently developed are used for materials having higher notch toughness (i.e., exhibiting elastic-plastic behavior). Numerically, the stress-intensity factor K has units of ksi $\cdot$ in$^{1/2}$ (MN/m$^{3/2}$). Equations for calculating the stress-intensity factor K have varying forms depending upon body configurations, crack sizes, orientations and shapes, and loading conditions. These are available from the fracture mechanics literature. For example, the stress-intensity factor for an infinitely wide plate subjected to a uniform tensile stress σ, and having a through-thickness crack of length $2a$ is $K_I = \sigma\sqrt{\pi a}$. For proper insight into a material's fracture characteristics and comparisons among materials, K_{Ic} values should be paired with corresponding yield strength values. Fracture toughness, the property of significance to crack propagation, is a function of $(K_{Ic}/\text{yield strength})^2$.

Figure 3.5 shows the general relationship of K_{Ic} to tensile loading and flaw size. Note that K_{Ic} is not a point but a locus of points—a curve. Also note that small flaws require high stress for unstable crack propagation, and that larger flaws can tolerate only lower stresses. Catastrophic fracture occurs when the flaw exceeds critical size at a constant stress level, or when stress on a stable crack of critical size increases. For avoidance of brittle fracture, tensile loading and crack size must be controlled within the indicated limits.

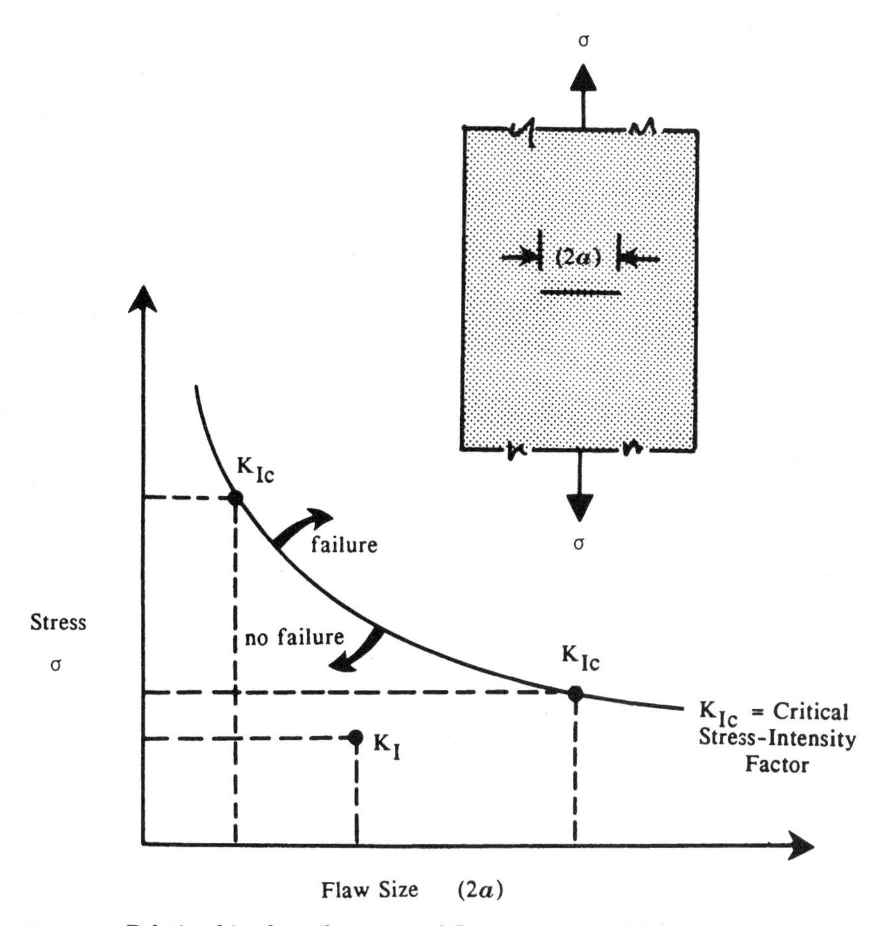

Figure 3.5 Relationship of tensile stress and flaw size to material fracture toughness.

K_{Ic} concepts founded upon linear-elastic or plane-strain conditions provide valuable insights into practices and procedures for avoiding brittle fracture. However, for plate materials of typical thicknesses used in most engineering applications, K_{Ic} determinations cannot be made due to the inability to achieve plane-strain conditions. Also, many engineering materials in current use fracture in the elastic-plastic mode, making K_{Ic} determinations impossible as a result of the formation of large plastic zones at the crack tip. In response to the need for analytical procedures for assessing fracture behavior in such materials, research effort has been under way for some years in attempts to devise procedures that will supplement or extend earlier work in the linear-elastic field.

Several approaches have been recently developed. The most promising of these have been (a) crack-tip opening displacement (CTOD) tests, (b) R-curve analysis, and (c) the J integral. Standardized procedures have now been established for these tests that evaluate materials' resistance to fracture during incremental slow-stable crack extension resulting from growth of the crack-tip plastic zone as the crack extends from a sharp notch.[51–53] These tests are useful in determining fracture characteristics for materials that are too ductile or too thin for valid K_{Ic} determinations.

Rate of loading can result in different fracture toughness values as, for example, during impact loading. Therefore, tests have been developed to measure K_{Ic} under dynamically loaded conditions, and is designated as K_{Id}. These have not been standardized, but most are some form of instrumented precracked impact-loaded bend test. The effect of loading rate is not observed in all materials, but this possibility should be investigated before using inappropriate test values.

Control of brittle fracture. The development of fracture mechanics concepts has been instrumental in providing essential baseline information on avoiding brittle fracture. In actual practice, however, it is often not feasible to conduct K_{Ic} (or even J_{Ic} or other similar tests) evaluations to determine suitability for specific designs and applications. As noted above, material size may not lend itself to such tests. Also, costs of conducting the required tests can be prohibitive. Much work has been done in devising auxiliary test procedures based upon fracture mechanics concepts and in developing empirical correlations to provide estimates of K_{Ic} for commonly used materials.[54] As a result, many of the notch-toughness tests mentioned earlier, with modifications based upon fracture mechanics principles, are being widely used in measuring a material's fracture resistance to various service temperatures and loading rates.

As things stand today, many useful tools are available to engineers for minimizing occurrences of brittle fracture. No design should be contemplated or materials application made without due consideration of these issues. The particular approach will depend upon the material, the application, its requirements and desired performance, the consequences of failure, and economics. These will largely dictate the fracture criteria used. There are no hard and fast rules for minimizing brittle failures in all situations; only basic guidelines and general principles.

Brittle fractures originate at defects. Cracks develop at defect sites and grow by various mechanisms such as fatigue, stress-corrosion, and hydrogen embrittlement. Crack growth requires tensile stresses oriented in a direction normal to the plane of crack propagation. Growth rate and fracture behavior of the component are functions of the material, service temperature, environmental effects, loading rate, section size of the component, and other variables.

Total elimination of defects to preclude brittle fracture is an impractical approach because of the impossibility of doing so, as all metal parts contain defects of some kind. The best and most thorough inspections cannot locate all of them, as some are beyond the detectability limits of the inspection method. Even if all original defects were capable of detection and elimination, other defects will be inevitably introduced during the lifetime of the component. These can become sites for crack initiation by any number of the mechanisms listed in Table 3.1 to which the material may be susceptible.

Notwithstanding this, good engineering practice dictates that defects should be minimized as a first step in avoiding brittle fracture. Large structures have failed by brittle fracture during initial proof testing, before seeing any service, and at low stress levels due to the presence of defects exceeding critical size at the time of construction, defects that were undetected earlier by radiography and other inspections.[55]

Brittle fractures have been traced to accidental welding arc strikes on vessel surfaces. In some steels, such incidents can produce a localized region of brittle untempered martensite (from rapid quench of small volumes of melted and resolidified metal) that can crack merely from cooling and/or transformation stresses. Similarly, sites of welds to attach temporary lifting lugs and instrument brackets have initiated catastrophic brittle fractures.

Catastrophic failure in March 1980 of the *Alexander Kielland,* a North Sea oil drilling rig, resulted in the loss of 123 lives. The failure origin was identified as a small fillet weld joining a non-load-bearing flange to one of the main bracings. The flange plate was used to support a sonar device.[56] Such welds are often an afterthought, not part of the overall construction, and can be made under adverse and largely uncontrolled conditions using inappropriate methods. This can result in hydrogen embrittlement and/or crack-prone untempered microstructures.

Other fracture-originating defects may arise from grinding marks, hammer dents, corrosion pits, stenciling of identifying marks with spark-erosion electric pens or marking stamps, and machining or flame-cutting operations carried out in the field using inappropriate tooling and procedures. Welds offer numerous opportunities for cracks and other fracture-initiating defects, and these are discussed later in a separate section.

Unexpectedly high levels of tensile stress can also contribute to the growth of cracks and flaws. These can arise around stress-concentrating sharp notches and material imperfections and in regions of abrupt changes in part geometry. Residual stresses in the vicinity of welds can be of yield-point magnitude, and material surrounding shrink-fit members can also contain high residual tensile stress. These are additive to operating stresses such that actual stress levels imposed upon cracks, flaws, and notches can be much higher than predicted from design calculations and cause crack growth rates that can seriously shorten component life. Temperature changes in components and assemblies containing materials of different expansion coefficients can introduce additional stress.

Also, along these lines, part size and thickness can increase constraint in the vicinity of flaws and defects and result in developing plane-strain conditions that limit plastic deformation in these regions and cause brittle fracture. This may occur even though similar structures or components of smaller size and thinner sections of the same material may not have behaved in this manner.

Choice of material can have a significant influence upon failure by brittle fracture. Materials susceptible to cracking during manufacturing or when exposed to expected service conditions and environments should be avoided. If this is not possible, steps should be taken to minimize it. If the component is to be constructed of an alloy that experiences temperature-dependent ductility transition, it should be ascertained through appropriate tests that the material in the final condition used for the component has adequate toughness at the service temperature. This holds for welds and all fittings and components that become part of the structure.

A good example of real-world complications that can lead to unexpected brittle fracture is the exploding stainless steel canister described in Chap. 1. Ordinarily, type 304 stainless steel is a ductile material, particularly at elevated temperatures. In fact, researchers have encountered difficulty in attempting to obtain valid K_{Ic} values for the alloy at ambient and lower temperatures due to its inability to develop plane-strain conditions in even heavy-section test specimens. Nevertheless, the severe sensitizing embrittlement that occurred in the canister provided a grain boundary path at least partially through the vessel wall that had brittle characteristics—properties that were not possessed by the original material. Under the right conditions, it failed by catastrophic brittle fracture (see Figs. 1.5 and 1.6) as surely as it would had the entire alloy been a brittle material of low K_{Ic}.

For high strength-to-weight applications, considerable vigilance is required for a number of reasons. One is that higher operating stresses diminish critical flaw size (recall Fig. 3.5) and may place it below the limit of detectability of standard inspection methods. In addition, original minute and undetected flaws can grow to critical size in service and escape detection until brittle fracture occurs. New flaws can also undergo the same process. Stress-concentrating effects and residual stresses will also be increased, further aggravating the higher overall design stress situation. Then, too, higher strength alloys are more prone to cracking-type degradation modes (e.g., stress-corrosion cracking and hydrogen embrittlement) and the crack susceptibilities of metallurgical transition products, embrittling second phases, and precipitates.

As material strengths increase, the need for close attention to the principal factors contributing to brittle fracture becomes increasingly critical. These factors include stresses being applied from operating loads as well as from stress-concentrating geometries. The list also includes notches and flaws; the size, shape, and locations of flaws; and the inherent toughness characteristics of the material in its final service condition. Projected service life also must be considered. Components needed for only brief intervals, as in missile delivery systems, would be less susceptible to problems of crack growth than components that must function without failure for extended periods of time. Vessels containing toxic substances under pressure, nuclear materials containment, and other high-risk applications will require stringent adherence to practices that will avoid brittle fracture.

Consequently, each application must be evaluated separately and thoroughly for fracture susceptibility and fitness for the intended purpose. Systems-analytical techniques described earlier can help to ensure that all relevant factors are considered, whether in the design stage or in operating equipment.

In the following discussions of crack-initiating failure mechanisms, the principles of avoiding brittle fracture apply throughout. These are cracking mechanisms, and response of material to the resulting cracks, of whatever origin, will be in accordance with these principles. The first two mechanisms discussed, stress-corrosion cracking and hydrogen embrittlement, are often referred to as *sustained load environmental cracking*. As many as 12 common characteristics have been identified for this category.[57] Liquid metal embrittlement[58] is often included, but is not discussed here as it is less frequently encountered. Embrittlement from neutron irradiation is another form of metal damage that has some similarity to the other three.

Stress corrosion

Nature and scope. This is a particularly damaging degradation mechanism. Under the most severe conditions of tensile loading, environment, and material susceptibility, it can cause failure of a part within minutes. Usually, failures take longer; but days, weeks, and months are not unusually short lifetimes under adverse conditions.

Materials affected by stress-corrosion cracking usually have no outward symptom of the fine cracks penetrating deeply into the component. As in brittle fracture, crack planes are oriented normal to the direction of applied tensile stress. External loads often are not needed for it to occur, as residual stresses from various sources can be sufficient for it to progress unhindered.

Its name is descriptive in that the attack is brought about by the simultaneous presence of tensile stress and a corroding environment. But it is a deceptive description as the combined effect is significantly more severe than that expected from the sum of the two conditions of corrosion and stress acting alone. Stress-corrosion cracking requires three simultaneous conditions: (a) a susceptible material, (b) tensile stress, and (c) an environment (usually aqueous) that triggers the attack in the given material.[59,60]

All metal systems are susceptible, although pure metals (elemental forms) have a high degree of immunity, if not complete immunity. Specific environments tend to single out particular alloy systems. The range of environments that causes a particular alloy system to crack is relatively narrow. For example, brasses and austenitic stainless steels (3XX series) are common materials that are susceptible to attack, but in different media. Brasses crack in ammonia environments and stainless steels do not, whereas stainless steels crack in chlorides, but brasses do not.

Among the first documented examples of this kind of attack were riveted mild steel boilers of early steam locomotives. They occasionally exploded from cracks originating at rivet holes, which had been stressed beyond the elastic limit by the expanding rivets. Boiler water had been treated at the time with caustic (sodium hydroxide) to control general corrosion. Evaporation of water seeping into crevices between rivets and the boiler plate concentrated the caustic within the crevices in the highly stressed regions. When concentrations were sufficient for stress corrosion to occur, cracks would proceed until leaks occurred or, if conditions of boiler temperature, stress, and steel fracture characteristics were right, catastrophic fracture from propagating cracks would cause the pressurized steam boilers to explode. This particular type of degradation of steels was termed *caustic embrittlement.*

Another early occurrence involved cracking of brass military cartridge cases in the region of the tapered crimp (deformed metal) that secured the projectile or bullet. The cracking appeared to be more prevalent during rainy seasons in tropical climates and became known as *season cracking.* The cause was eventually traced to ammonia from decomposing organic matter generated by warm and moist environments.

Various conditions, such as increasing temperature and level of stress, can aggravate stress-corrosion and shorten the time to failure (i.e., accelerate crack growth). Other variables are metal composition, alloy heterogeneity and microstructure, and its heat-treated or mechanical condition. Composition and concentration of the corrosive aqueous solution are other factors.

Stress-corrosion cracks resemble brittle fractures as they are usually flat and granular appearing with little or no accompanying plastic deformation. And, like brittle fractures, crack planes are oriented normal to the applied tensile stress. In some materials and under some conditions the cracks tend to be branched. This can be a clue in identifying the fracture mechanism as stress-corrosion, but it is not a wholly reliable indication as some stress-corrosion cracks are not branched. Cracking can be intergranular or transgranular, depending upon the material and its metallurgical condition, the environment, and other factors.

Figure 3.6 shows the appearance of crack profiles and a fracture surface of a failed aircraft landing gear component of a 7XXX type aluminum alloy. Brittle fracture initiated from stress-corrosion cracks along the inner surface of a hole adjacent to a highly stressed pinned connection. The upper edge of the part as shown in the fracture surface photograph is the location of the stress-corrosion. The cracks were intergranular and believed to have been initiated by moisture seeping along the interface between the steel pin and aluminum component.

When cracks had progressed to the lower side of the ragged-appearing upper band (numerous crack fronts, probably of critical size for this alloy), their crack fronts apparently merged into a single crack that propagated by brittle fracture across the entire component. This failure scenario is typical of many stress-corrosion cracking occurrences, where intermittent stress-corrosion, progressing when moisture was present and stress conditions were right, set the stage for eventual brittle fracture.

Contributing factors. The stress-corrosion cracking phenomenon is not completely understood, although progress in its understanding is being made.[61] The initiating site may be almost any surface feature or artifact that is a discontinuity or macrofeature that serves to concentrate tensile stresses, rupture surface films, or concentrate the corroding medium. Surface pits, scratches, gouges, and dents are common sites. Corrosion at the crack tip appears to contribute to the phenomenon as experiments with cathodic protection superimposed upon an advancing stress-corrosion crack stopped its progress despite continued application of the tensile load. Cracking proceeded once again when the protective cathodic current was removed.[62]

It is believed that tensile stresses imposed at the crack tip promote the rupturing of protective films that tend to form on many alloys in ambient environments and aqueous solutions. This is thought to occur both during the initiation stage as well as during crack propagation. Film rupturing at the crack front would permit active corrosion to proceed into fresh unprotected metal followed by formation of corrosion

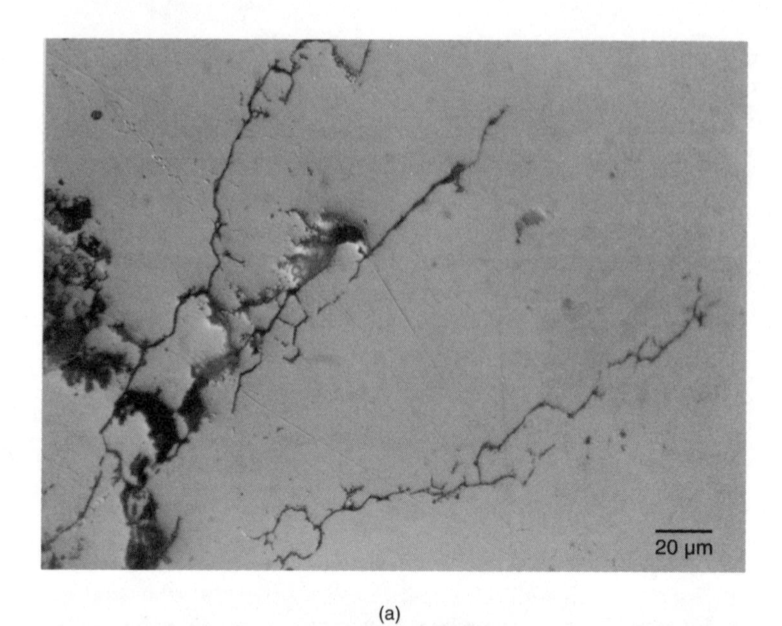

(a)

(b)

Figure 3.6 Fracture in 7XXX aluminum alloy initiated by stress-corrosion cracking (showing unetched cross section and fracture appearance).

product that would tend to stifle attack and perhaps create a microwedging effect at the crack tip. Continued application of tensile stress would, however, rupture the protective layer, allowing the attack to once again proceed. It is also believed that the presence of hydrogen at the crack, generated by the corrosion process, plays a role in the cracking process and its progression.

Crack paths through grain boundaries might allow faster propagation because of less resistance or more favorable corrosive conditions, and this is often observed. An example is stress-corrosion attack in sensitized stainless steels, where grain boundary regions have been depleted in chromium and therefore are anodic to the surrounding matrix.

If the above descriptions are an accurate depiction of the stress-corrosion process, it might be expected that any superimposed cyclic stress generated from local load fluctuations or mechanical vibrations would speed up the film rupturing and crack-tip progression, and this has been the experience.[63] Small-amplitude flutter superimposed upon a mean sustained tensile load was found to cause a particularly damaging stress-corrosion situation. This condition might be developed in the vicinity of a pump station or pressurized pipeline. Under such conditions threshold stress for stress-corrosion initiation decreases significantly.

The average concentration of corrodent is not a reliable guideline in assessing material susceptibility in a given application or environment. Threshold concentrations (for onset of attack) can be readily exceeded through localized evaporative losses of water, tending to produce surface residues that become gradually enriched in the corroding species until a sufficient concentration is present to initiate attack. Alternative wetting and drying often cause more severe attack than exposures to single-phase (constant immersion) environments.

Cracks have occurred in materials subject to rainwater accumulations. Rainwater washes over surfaces exposed to atmospheric contaminants and can run into and accumulate within crevices where the water evaporates, concentrating the contaminant. Repetitive cycles can lead to concentrations that exceed threshold values for attack to initiate. A failure of 0.7% carbon-content steel bridge cables was traced to residues of ammonium nitrate from the atmosphere that had washed into crevices at cable attachment locations by rainwater and had accumulated there and become concentrated.[64]

Similar occurrences have been noted when rainwater and other sources of moisture leached corroding impurities from nonmetallic materials above or in contact with susceptible materials. Pipe insulation, concrete, gasket and packing material, and lubricants have been sources for such corroding impurities.[65] Harmful substances have also been leached from plastics and other organics that had been thermally decomposed and from adhesives, binders, and other fibers and constituents. Alternating wetting and drying during exposure to tides and chemical process effluents can have the same effect.

Often, only relatively low concentrations are necessary for initiating and sustaining attack. Concentrated solutions are not necessary; the mere presence of trace amounts of the corroding ions can cause cracking. In fact, plain water is sufficient to initiate and sustain stress-corrosion in some high-strength aluminum alloys.[66] Austenitic stainless steels (3XX series), even in the solution heat-treated or annealed condition, have cracked in 200°C (390°F) aqueous solutions that contain as little as 2 parts per million (ppm) of chloride. The same steels in the sensitized condition crack at room temperature in water containing as little as 10 ppm of chloride or 2 ppm of fluoride.[67] Similarly low concentrations of ammonia can crack some copper alloys.

The writer observed rapid attack some years ago in high-strength steel test weldments that had been shipped inside wooden crates to a testing facility located on the sea coast only 300 miles away. Upon opening the crate a few days later, the testing laboratory found the samples cracked into fragments inside the packing material within the sealed crate. This occurrence was traced to chlorides from the marine atmosphere that had penetrated the crate and packing material. These trace residues had been sufficient in the warm and humid climate to attack and fragment the highly susceptible steel weldments (containing high residual stresses) and accomplish it within a few days.

There was speculation that hydrogen embrittlement may have been the dominant factor in this incident. However, precautions had been taken to minimize this; besides, similar samples from the same welds that had been retained at the laboratory and not exposed to marine atmospheres remained intact. It is likely, however, that hydrogen embrittlement played a role, as it probably does in virtually all stress-corrosion cracking.

As indicated, continuous immersion in aqueous solutions is not always necessary, as occasional moisture may be all that is required. And the problem is not confined to heavily loaded members, as residual stress from forming operations, welding, shrink-fit parts, differential expansion and contraction, and other sources may be sufficient. It is advisable to regard all materials as susceptible to some environment. Susceptibilities of common engineering materials and their respective environments are well documented in the literature.[68–70]

For critical applications, unknown environments, and nonstandard material compositions, forms, or conditions, there is no effective substitute for conducting actual tests. There are numerous standardized procedures for doing so.[71–78] Specimens should be carefully prepared and the tests conducted to avoid extraneous contaminants that could affect results. Simulated service tests do not usually duplicate actual conditions and results must be interpreted accordingly.

A major problem is accommodating the time factor, as laboratory tests usually must be of relatively short duration whereas failures in actual environments may take considerable time, depending upon stress fluctuations, environmental concentrations, susceptibilities, and many other variables. Time-to-failure laboratory tests must, therefore, be used strictly as basic guidelines for making decisions. This requires considerable experience in the field of stress-corrosion and sound understanding of the materials, environment, processes, and exposures involved.

Relevance of fracture mechanics principles. In recent years, attempts have been made to apply fracture mechanics principles to stress-corrosion cracking to provide more quantitative information and guidance in avoiding this kind of attack. Stress-corrosion cracking is really one type of brittle fracture; as it occurs at stress levels well below the material's yield strength, it is dependent upon tensile stresses for its initiation and propagation and occurs with little or no accompanying plastic deformation. Therefore, there would appear to be a logical similarity for these attempts.

For the most part, they have been successful. These studies have concentrated upon using the stress-intensity factor K_I to characterize the mechanical component of the driving force in stress-corrosion cracking.[79] Time-to-failure and crack-growth-rate tests have been devised using various precracked specimens and loading methods. Using these tests and the concept of establishing a K_{ISCC} threshold (maximum stress-intensity factor for avoiding stress-corrosion cracking) for a given material–environment combination, results have demonstrated the feasibility of identifying a plane-strain stress-intensity factor below which statically loaded structural components are expected to have infinite life when subjected to the specific test environment.

Continuing work suggests that K_{ISCC}, like K_{Ic}, may be regarded as a material property and, therefore, affords a basis for assessing the reliability and integrity of engineering structures. A drawback is that standardized test procedures have not yet been established. This probably will not occur until the effects of various critical test parameters have been reconciled. Specimen size and test duration are two primary considerations, and there are others.[80] Until standardized procedures have been established for determining K_{ISCC}, it is advisable to avoid using published values of this parameter as the basis for design decisions. However, such values may be useful in evaluating relative susceptibilities.

Material variables. It is important to realize in evaluating reported mechanical property and fracture toughness data for metallic materials, including indicated thresholds such as K_{ISCC}, that most of these materials are not isotropic, that is, their properties can differ significantly with testing direction. Furthermore, there is usually a size effect that is related to microstructural homogeneity, grain size, and grain shape and orientation. These variables are introduced during mechanical processing of metals and alloys from their original ingots (via forging, cogging, rolling, etc.). To some extent, they can also be introduced by the solidification behavior of the particular alloy system: how it was cast, the ingot shape, pouring temperature, and pouring and casting atmosphere. Metal purity is also a factor. Alloys melted from high-purity melt stocks under vacuum will behave differently from those using scrap charges and melted in air. Melting furnaces, melting processes, and melt-treatment practices are other variables.

These can have a significant influence upon damage susceptibility and response of structural members and components fabricated from metallic materials. For example, fracture toughness determined in specimens where crack propagation is normal to grain textural direction can be markedly higher than when it is parallel to the texture. Such *anisotropy* is evident in other structure-dependent properties, including those related to stress-corrosion cracking and fatigue. Therefore, for critical applications, as well as others, it is prudent to take the time to determine how properties were derived and what material condition and testing direction were used, before applying them in either design calculations or life-extension determinations.

Control measures. Despite the stress-corrosion cracking susceptibility of virtually all engineering alloys to some environment or conditions, most structures, components, and hardware systems are operating without serious risks of failure by this mechanism. This is possible because precautions are taken to avoid these failures, and because susceptible combinations fail and force engineers and users to consider alternative materials or to make design or process modifications to avoid recurrences.

Avoidance measures are developed around knowledge of causation elements, and these—as stated at the beginning of this discussion—are the simultaneous presence of a susceptible material, tensile loading, and an environment that attacks the material. One step might be to decrease the operating stress to below the threshold for stress-corrosion crack initiation, provided that a stress threshold exists for that alloy (some aluminum alloys, for example, do not exhibit one). Significant relief might be achieved through minimizing residual stresses through stress-relief annealing and eliminating regions of stress concentration. Since stress-corrosion requires *tensile* stress, it is frequently helpful to introduce compressive stresses at the surface (where stress-corrosion cracking initiates) using surface peening or shot blasting. This treatment can work as long as compressively stressed layers remain continuous and intact and are not removed by erosion, general corrosion, or other conditions.

If feasible, elimination of the corroding species will remove one of the primary causal elements and stop the attack. Decreasing temperature can also reduce susceptibility. Concentrations of evaporative residues may be eliminated through process or vessel modifications to ensure against repetitive wetting and drying cycles in splash or tidal zones. Elimination of process leaks and drips onto susceptible materials should also help. Knowledge of the existence of leachable species from surrounding materials and their elimination can also help to avoid stress-corrosion cracking incidents. Failures should be thoroughly investigated to identify the responsible species and its source. These should be eliminated before operation is resumed with replacement components.

Since it has been demonstrated that stress-corrosion crack growth can be stopped by applying cathodic protection, this is a possible approach for controlling attack. It is important to realize, however, that cathodic currents generate hydrogen at the protected metal surface. If the material is sensitive to hydrogen embrittlement, as are many alloys susceptible to stress-corrosion, the "cure" may lead to a worse situation than the original problem. Also, apparent similarities between hydrogen embrittlement and stress-corrosion cracking make this approach risky unless it has been conclusively established that the problem being corrected is truly stress-corrosion cracking.

Corrosion inhibitors are sometimes added to the corroding environment to assist in minimizing this problem. Their use and effectiveness in the specific combination of interest must be established, along with the necessary concentrations, monitoring procedures, and makeup intervals.

For some highly susceptible alloys, such as high-strength aluminum alloys used in aircraft construction, surface cladding of leaner alloy content (and, therefore, less susceptible to attack) is often applied at the mill to plate and sheet forms to impart resistance. These are standard product forms commonly listed in suppliers' catalogs.

It is evident that stress-corrosion cracking is a major consideration in avoiding component failures. It is a common fracture mechanism and should be included in every assessment of failure potential in hazards analysis, FMEA, and FTA.

Hydrogen embrittlement

General characteristics. This fracture mechanism, second of the two examples of sustained-load environmental cracking, has many similarities to stress-corrosion cracking, the other example described above. Both cause brittle fractures, although susceptible alloys may exhibit ductile behavior in the usual uniaxial tensile test. Cracking in both occurs at stresses well below the material's yield strength. These fractures occur only under tensile stress. The environmental species that trigger the cracking need not be present in large quantities or high concentrations. There are many other similarities, including their time dependency for fracture.[81]

Despite the many similarities between stress-corrosion cracking and hydrogen embrittlement, they are not the same. This is demonstrated in experiments on materials where cathodic protection initiates cracking under the superimposed electric current (which generates hydrogen at the metal surface). In materials subject to stress-corrosion but immune to hydrogen embrittlement, cathodic current stifles crack progression.

Hydrogen embrittlement, sometimes referred to as hydrogen-induced fracture or hydrogen-stress cracking, is but one form of metal damage caused by hydrogen. However, all reduce the load-carrying capacity of the metal or alloy when accompanied by tensile stress. The most common form of hydrogen damage typically is encountered in high-strength steels of tensile strengths above about 145 ksi (1000 MPa), characterized by marked loss of tensile ductility under slow-strain rate conditions. This is a result of internal absorption of, or penetration by, hydrogen before external stress is applied.

Embrittlement is observed at room temperature as a delayed catastrophic brittle fracture at stress levels far below the yield, or nominal design strength, for the alloy. The delay can be any interval ranging from minutes to years after a static load is applied to the hydrogen-containing structure or component. Embrittlement of lesser severity may be observed as ductility loss in tensile testing under slow-strain-rate conditions.

Cracks initiate internally, usually below the root of a notch in regions of maximum triaxiality.[82] Embrittlement in steel is reversible (i.e., ductility can be restored), provided that microcracks have not yet formed, through relieving the applied stress and holding at room temperature or baking at temperatures in the vicinity of 200°C (392°F). It is the nascent (atomic) form of hydrogen that is responsible, as hydrogen atoms have a high degree of mobility and can readily diffuse through the metal lattice and collect at internal microvoids, crystal lattice defects, inclusions, and, at elevated temperatures, grain boundaries.

Three classes of hydrogen-induced damage have been identified: (a) internal reversible hydrogen embrittlement, (b) hydrogen environment embrittlement, and (c) hydrogen reaction embrittlement.[83] The first, briefly described above as simply hydrogen embrittlement, has received the most attention as it has probably accounted for most of the catastrophic hydrogen-induced failures. In these failures, atomic hydrogen from any number of possible sources and reactions diffuses into the metal and causes delayed cracking.

The second class of damage is experienced in hydrogen containment or storage vessels subjected to pressurized gaseous hydrogen. There has been some dispute over whether this form of damage is simply a secondary manifestation or another form of the first class.[84]

The third type of damage involves chemical reactions between hydrogen and the metal itself, or some constituent of the metal or alloy, to form embrittling species. The reaction may take place at the surface, or the hydrogen may diffuse some distance into the metal before the reaction occurs. The reaction product may be a hydride of the matrix metal or one of its constituents (e.g., as in titanium, tantalum, or zirconium).[85,86] The presence of hydrides increases strength at the expense of ductility and toughness and degrades the usefulness of affected materials.[87]

Another internal hydrogen reaction can occur when diffused hydrogen atoms recombine to form molecules of hydrogen (gaseous form) causing internal pressure and local decohesion. Or, the hydrogen may react with other elements within the metal to form a gas. It combines with carbon in low-alloy steels at elevated temperatures to form methane ("hydrogen attack"). Methane bubbles form along grain boundaries and tend to agglomerate and form internal voids.[88] These weaken the metal from internal decarburization and can reduce fatigue life, stress-rupture, and creep ductility. A somewhat similar reaction can also occur in nondeoxidized (i.e., tough-pitch) copper when absorbed hydrogen reduces internal particles of copper oxide to form pockets of water vapor. These expand and cause internal blistering and fissures.

Our primary concern here, however, is with the first class of hydrogen damage, as this mode can cause catastrophic failure with no prior symptoms or advance warning, and it accounts for most of the hydrogen-induced damage experienced in industry. A petrochemical industry corrosion specialist reported a few years ago that stress-corrosion cracking and hydrogen embrittlement alone accounted for more than 50% of all failures experienced by one major oil company.[89] Although hydrogen embrittlement has been observed in other alloy systems, its effects upon medium- and high-strength steels are so severe and widespread that we will confine our attention to this class of materials.

Prerequisites for occurrence. The general consensus is that three conditions must be satisfied for occurrence of hydrogen embrittlement: The steel must have a heat-treated strength level above some threshold value; 100 ksi (690 MPa) is often given as an appropriate value, although it should not be considered absolute for all situations. Susceptibility increases with increasing strength level, and the steel must also be under tensile load, the threshold level depending upon the strength of the steel. Also, the steel must contain diffusible hydrogen.[90]

Threshold hardness or strength level for hydrogen embrittlement also seems to depend upon environmental conditions and other variables. For example, steels having heat-treated hardnesses below 22 Rockwell C (Rc), corresponding roughly to a tensile strength level of 116 ksi (or 800 MPa), are considered satisfactory for the oil industry use where sulfide exposure is involved.[91] Embrittlement in ferritic steels occurs within the range of −70 to +140°C (−94 to +284°F) and is generally most severe at room temperature. The cracking tendency diminishes with increasing temperature such that in steels it disappears at temperatures above 200°C (390°F).

Room temperature (20°C or 68°F) embrittlement under applied tensile stress is observed in steels containing as little as 0.1 ppm of absorbed hydrogen. Cracking tendency increases with increasing hydrogen content such that at a level of about 10 ppm, cracks can form spontaneously without externally applied stress. Obviously, the presence of residual stresses or stress-concentrating factors plays a major role in the occurrence of brittle fractures due to absorbed hydrogen, particularly in steels of high strength levels and correspondingly high susceptibilities. Hydrogen-induced fracture is favored by external tensile loads that are static or increase slowly. Internal stresses, such as those imposed by contraction and expansion, metallurgical transformations, and conditions that increase hardness and strength, such as aging, are the kind that lead to cracking in the presence of diffusible hydrogen.

Hydrogen can come from any source; the range of possibilities is wide. Electroplating processes are a common source in embrittlement of steels. These processes include cathodic cleaning, pickling, activation, and plating steps, as hydrogen is a by-product of reactions that take place during these processes. Although most of the hydrogen formed during these operations is harmlessly released to the atmosphere as molecular hydrogen, small volumes of dissociated (atomic) hydrogen enter the metal. It is this form of hydrogen that readily penetrates the metal and causes embrittlement.[92] The problem with electrochemically inoculated hydrogen is particularly acute in aerospace, aircraft, and other industries where high-strength components are electroplated to protect against corrosion and wear.

Corrosion is a galvanic (electrochemical) process and can inoculate members of the corrosion couple with hydrogen, as it is a product of the reaction. Figure 3.7 shows the appearance of a brittle fracture of an unprotected high-strength steel bolt (200 ksi, or 1380 MPa, tensile strength) that occurred by hydrogen embrittlement. The bolt had been under tension while in direct contact with an aluminum component. In

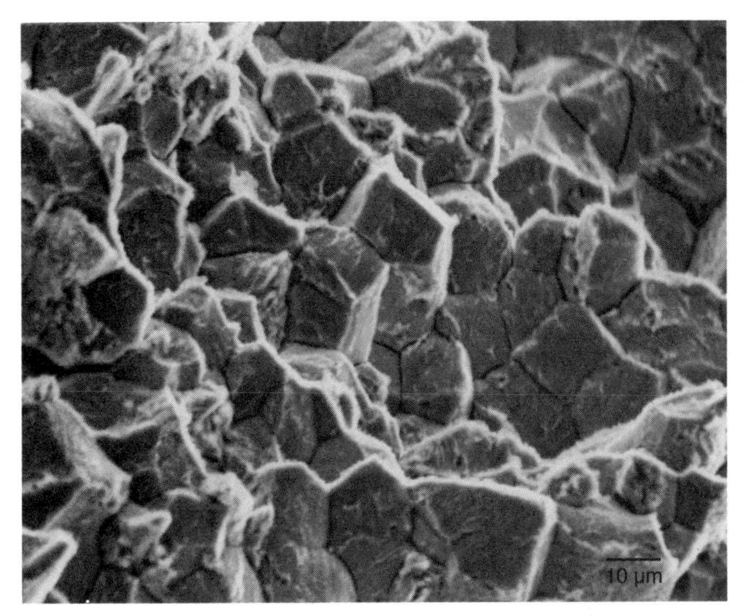

Figure 3.7 Brittle fracture of steel bolt as a result of hydrogen embrittlement caused by galvanic corrosion (scanning electron photomicrograph).

the presence of atmospheric moisture that had apparently seeped along the juncture of the dissimilar metal couple, corrosion had occurred. Hydrogen generated during the corrosion process entered the steel bolt, which was under static tensile load, causing its embrittlement and fracture.

Hydrogen charging of a high-strength steel part has also been observed during electrochemical-discharge machining (EDM). In this spark-erosion process hydrogen is a by-product of the arcing occurring within the electrolyte bath at the surface of the part. Atmospheric pollutants deposited on metal surfaces can affect the absorption of hydrogen into steels and other metals in the presence of moisture through a localized corrosion + hydrogen dissociation process. Cracking of steel tubes showing evidence of brittle fracture was traced to moist environments containing hydrogen sulfide and was believed to have occurred by hydrogen embrittlement (sometimes referred to as sulfide-stress cracking, or SSC).[93,94]

Cathodic protection using sacrificial anodes or superimposed current, mentioned previously, is commonly used for controlling corrosion of marine structures and buried pipelines. The protected metals are not affected by released hydrogen as the selected materials are not susceptible to hydrogen embrittlement. However, high-strength steels in the vicinity or within the protective circuit could become embrittled.

Atmospheric corrosion is another possible source of hydrogen in embrittling high-strength steels under tensile stress. The chemical reactions responsible for crack initiation and propagation are complex and have been the subject of considerable research. Moisture is necessary for galvanic corrosion to occur and is the source of hydrogen via electrolysis, even though it operates on a microscale. Hydrogen embrittlement requires only minute amounts of diffusible hydrogen to initiate cracks and sustain their propagation, and its source is unimportant. Once a corrosion couple is established in a susceptible material, a supply of hydrogen is available. If the part is subjected to an applied tensile stress, an initiated crack will progress. Fresh metal at the crack tip becomes available as the crack opens and this sustains the corrosion process, thereby maintaining the hydrogen supply for continuing the crack.

The hydrogen sources mentioned above are some of the possibilities for processed steel in finished forms. However, hydrogen can be readily introduced during the steelmaking process itself and cause problems at an early stage. This stems from the much higher solubility of hydrogen in steel in its liquid state. If cooling rates are too high for hydrogen to diffuse out during cooling, it becomes trapped within the metal lattice and can cause internal cracking of ingots and flaking in heavy sections during subsequent processing. The cracking will probably be delayed during some incubation interval (days or weeks).

Various degassing methods, such as establishing a carbon monoxide boil during melting or use of ladle degassing treatments upon pouring, are available for minimizing hydrogen damage in the mill. However, the same problems and the same general reasons for these problems exist in welds and can have even more serious consequences. Hydrogen-induced cracking is a potential problem whenever high-strength steels are welded. These problems and their control will be covered in the section dealing with welds as a favored site for fractures and their initiation.

Evaluation and avoidance. Avoidance of hydrogen embrittlement in structures and manufactured components requires knowledge of susceptibilities and effects of service environments. Literally thousands of research programs have studied the problem and numerous tests have been devised to assess relative material susceptibilities, suitability, and response under given conditions. Obviously, no single test method can begin to fulfill the range of needs. A wide array of materials have been identified as having some degree of susceptibility to hydrogen damage.[95] Within each group are materials of varying degrees of susceptibility, depending upon their condition, strength level, and hydrogen content. Environments can affect materials' susceptibilities and response, and each application has its own set of requirements, risks, and performance expectations. Then there are the different modes of hydrogen damage to contend with.

In any test using reduced-scale specimens designed to provide information applicable to larger-scale components, it is always difficult to extrapolate results to actual situations. In addition to the matter of scale, there are the questions of whether the actual stress state has been duplicated and whether the specimen loading mode represents the real-life application, whether the test environment is severe enough or too severe, and what will be the effects of compositional differences and levels of interstitials, inclusions, and microstructural differences from test specimen to actual part. It is evident that it is not an easy problem to handle. There are no "go" or "no-go" indicators and no straightforward approaches for assessing hydrogen embrittlement susceptibility, either in new designs or in existing operating equipment.

Many types of tests and specimen configurations have been and are being used. Most are some form of sustained-load (static-load) test, slow-strain-rate, tension, torsion, or bend test. Some employ fracture mechanics specimens or modifications. The tests are run in various electrolytes developed for providing repeatable and controlled conditions of varying degrees of severity for hydrogen charging of test specimens. To limit the number of test variables and possible differences between test conditions and service environments, tests can be carried out under actual service conditions.

Some test specimens are purposely precracked (by fatigue), as done in fracture mechanics testing, to make the test duration more predictable. However, such tests only provide crack-growth information, which is usually of primary interest. Various tests have also been devised for evaluating embrittling effects of exposures to pressurized gaseous hydrogen, such as disk-pressure tests.[96,97]

Often, hydrogen embrittlement tests and related performance criteria are developed by major users of high-strength materials, such as the military and aerospace, aircraft, and petrochemical industries, and these become part of the performance specifications for purchased and supplied components. The list of MIL (Department of Defense) standards and specifications for electroplated high-strength parts and components, alone, is quite lengthy.[98]

Avoidance practices take several approaches. The most direct approach to avoiding hydrogen embrittlement is, of course, to use lower strength material that is not susceptible. A related technique is to modify the heat treatment to impart a softer temper (using higher-temperature tempering treatments). This may be useful if the lower strength (but higher toughness) is acceptable and if the tempering treatment does not introduce temper embrittlement or unfavorable microstructures.

Decreasing the stress imposed upon the part would appear to be effective, provided that it is feasible. This involves more than decreasing the load on the part and means accounting for and controlling residual stresses, stress-concentrating notches, defects, and part geometries. Surface treatments to impart compressive stresses, such as shot peening, can also be helpful if the worked layer remains intact and local tensile stress conditions do not counteract and supersede the surface effects.

Another method is to minimize operations that inoculate with hydrogen. In the author's experience during the early days of the development of maraging steels (250 to 300 ksi or 1378 to 2068 MPa yield strength) virtually all experimental welds $^1/_2$ in (13 mm) and thicker were found to be subject to delayed cracking, typical of hydrogen embrittlement. The welding operation, using inert gas–shielded processes, was scrutinized in attempts to identify the source of hydrogen. The shielding gas moisture content (dew point) was measured, all coolant and gas lines were replaced, plate materials were scrupulously vapor-degreased, cleaned, and dried, room humidity was controlled, and handling of all materials was minimized to avoid moisture pickup. Still, the cracking occurred—sometimes overnight or a day or so after completing the welds.

Then, someone questioned the filler wire as being the source. This was confirmed to be so when discussions with wire mill personnel revealed that it had been standard mill practice to copper-electroplate all difficult-to-draw wires, and these compositions certainly qualified. The electroplating operation involved electrolytic precleaning, electroplating, wire drawing, annealing, followed by repeated cleaning, plating, etc. The copper had been pickled off at the mill prior to delivery. Both final plating operation and pickling treatment had charged the weld filler wire with high levels of hydrogen. This transferred readily across the welding arc into the molten iron-base alloy weld where hydrogen solubility was high. Under conditions of rapid cooling, typical of weld metal, plus with the presence of high residual stress from the cooling and contracting weld deposits and the added mechanical restraint of weld fixturing, conditions were ideal for hydrogen-induced delayed cracking.

For electroplating operations, modifications that may assist in minimizing hydrogen embrittlement would be to avoid cathodic cleaning, pickling, or activation treatments and substitute anodic cleaning, etching, or electropolishing. Mechanical processes for scale removal can be substituted for pickling. And plating methods that avoid or minimize hydrogen inoculation, such as vacuum-vapor deposition, organic coatings, or modified electroplating processes that minimize hydrogen embrittlement, may be helpful.

Where electroplating is essential, parts should be heat-treated (baked) soon after plating to remove the hydrogen. This step is called out in many specifications for plating of high-strength components and the recommended procedures should be carefully followed. Post-plating baking treatments are not a panacea, however, as some electroplated surface deposits can be a barrier to hydrogen removal. A number of incidents of hydrogen embrittlement have occurred in electroplated components that had been given post-plating baking treatments. Because of this possibility, specifications usually call for appropriate tests and inspections on completed parts.

Despite the attention it has received from materials researchers and users and the need for eliminating these kinds of failures, hydrogen embrittlement remains a serious problem. The triggering element, hydrogen, is found everywhere as a by-product of reactions with water. These reactions can occur wherever water and metals come together. The hydrogen atom is extremely small and can readily diffuse through solid metals, virtually at will. It is a natural misfortune, indeed, that the greatest susceptibility for hydrogen embrittlement occurs at room temperature. Its damage singles out as its prime targets our finest achievements—high-strength alloys—intended to provide engineering efficiencies and structural advantages of high strength-to-weight materials. Aviation and aerospace technology, as we know them, could not exist without such materials.

The damage does not stop there, as almost all metals and alloys can be threatened, although under specific environmental conditions, with catastrophic brittle fracture occurring at strength levels far below textbook design values. The problem will not go away, but may worsen. Engineering progress is tending toward an increasing threat from this fracture mechanism as performance demands require yet higher-strength materials having greater endurance and dependability.

To achieve this, alloys and materials combinations are becoming increasingly complex. For higher efficiencies, mechanical equipment must operate longer, faster, under higher stress, more effectively, and under more adverse conditions than ever before. New processes require higher temperatures and pressures and involve more aggressive environments. All these are factors that increase the likelihood of the kinds of failures we have been discussing. Above it all, the costs of failures have skyrocketed as a result of developments in products-liability law, increased public concern and intolerance for industrial mishaps, accidents, disasters, and all technically related failures. Environmental, political, and economic constraints that largely dictate corporate policy tend to make society a business partner. Increased vigilance and a greater understanding of the approaches for minimizing failures are the only answer to avoiding them—not only of metal components and engineered systems, but of our engineering careers and businesses as well.

Fatigue

Cyclic-loading effects. By now, it should be evident that the damage mechanisms selected for discussion have much in common. All can lead to catastrophic failure. All occur with little or no apparent plastic deformation. All seem to abrogate traditional notions of the strength of materials in producing total fracture of a metal part at applied stresses well below its yield strength. All occur in normal service under usual operating conditions. Traditional safety factors used to calculate design loads and presumably sufficient to compensate for unknowns, inevitable flaws, and unforeseen consequences are inadequate to prevent these kinds of failures. And all, to some degree, are time dependent; that is, there is delay between the time of applying stress and ultimate failure.

Fatigue, our last example, has characteristics similar to the others. This fracture mechanism is said to cause or be implicated in more than 80% of all mechanical failures.[99,100] And it is probably the most difficult to deal with, as virtually all metals and alloys are vulnerable. Its "cause" is repetitive application of stresses and strains, over a wide range of temperatures up to, in fact, the melting point. Since the primary purpose of metal components is to carry and transmit mechanical loads, this constitutes a formidable and universal problem.

Its name, *fatigue,* or *metal* fatigue, as sometimes qualified, does not adequately convey its true character. This name was coined over a century and a half ago when metal parts—namely, railway axles—broke after extended periods of use and which, up to the time of their failure, had been altogether satisfactory. It seemed to early observers that they just became weary of carrying the load and, like people and horses, simply reached a point when they gave out, tired of it all. Laypersons, familiar with frequent references to the term *fatigue* in connection with metal fractures, still tend to view these failures this way. The enigma of why a component that had for some time successfully sustained a given stress would suddenly fail under its continued application has occupied the attention of metallurgists and engineers for well over a century.

In the earlier discussion of brittle fracture, we noted that subcritical cracks in materials do not extend without application of tensile stress across their faces—in the direction of separating the pages of a book (i.e., mode I). This stress, focused at the crack tip, is the driving force for crack progression. These variables depend upon the properties of the metal and intensity of the stress field ahead of the crack (a function of crack geometry). However, for a set of conditions where the stress intensity K_I is less than the critical value (K_{Ic}), the crack is stable, that is, it does not propagate.

In fatigue, however, if the stress is made to fluctuate above and below some median value—say, a complete reversal about zero—the cumulative effect of the fluctuations is propagation of the crack. In other words, stress *cycling* has accomplished what static stress imposed upon the crack did not. Somehow, stress fluctuations applied at the crack tip affect the material in that region in decreasing its resistance to propagation, if only by a small increment at a time. Of course, if load fluctuations are permitted to continue, the crack will grow to critical size (K_{Ic}) at which point crack progression becomes unstable and will suddenly proceed completely through the material, or at least far enough for the section to fail by tensile overload.

Even if fluctuating loads are applied to *uncracked* material, surface flaws, notches, or other stress-concentrating discontinuities can become sites for crack initiation. The result is the same as before (in the presence of a preexisting crack), except it will probably take longer for failure to occur because of the time required for crack initiation or incubation.

Fatigue cracking occurs in three distinct stages: (1) initiation (or nucleation), (2) propagation, and (3) final fracture. Fatigue produces a progressive, highly localized, but permanent structural change in materials and is brought about by the combined action of cyclic tensile stress and accompanying plastic strain.[101] To place this in real-life perspective, investigations have established that for most, if not all, metals and alloys, tensile loads that produce 1% plastic strain in the part or component can cause fatigue failure in about 1000 cycles.[102]

Crack initiation. Initiation in stage 1 is probably the most significant stage and the most complex. Cracks can nucleate at a free surface or below the surface. Most fatigue cracking involved in mechanical failures initiates at the surface, and our discussion will focus upon this location. Free surfaces are particularly vulnerable because of their exposure to the atmosphere, and this plays a role in at least some incidents. Surface grains are not entirely supported by adjoining grains and, therefore, can deform more readily.

The key role that the surface, with its susceptibilities to crack nucleation, plays is demonstrated in fatigue tests in which thin surface layers were removed after some fraction of the specimen's fatigue lifespan and the test resumed. Since it is the outer surface that experiences irreversible structural changes (*cumulative damage*), removal effectively restores the specimen to its original state, thereby extending fatigue life.

Although other mechanisms are also involved, studies of otherwise defect-free surfaces undergoing fatigue crack initiation have indicated a preference for nucleation at slip steps that form at the surface. Cracks nucleating at slip steps, or bands, progress inward at roughly 45° to the axis of the applied stress in a crystallographically shear mode for a depth of only a few grains (a few tenths of a millimeter). Observations made over the course of many experiments and for most metals and alloys have shown that such slip begins at a common value of critical resolved shear stress of about 200 psi (1.4 MPa) and a common strain of 10^{-4}. This point is considered the end of stage 1 and the onset of stage 2.

At low stress amplitudes, crack initiation can constitute a major portion of the fatigue life of a part, and techniques that delay crack initiation or prolong it can increase fatigue life. Treatments to restrict surface plastic deformation have been effective. Methods have included imparting compressive stresses through shot peening and surface rolling. For example, parts containing "rolled" threads have longer fatigue lives than those with machined threads. Surface alloying, as in steel carburizing, can have the same beneficial effect. Melting and steelmaking processes that produce cleaner material with fewer inclusions and defects provide fewer surface discontinuities and locations for crack initiation.

Crack propagation. Crack propagation (stage 2) is considered to begin as the crack front changes its orientation to that of normal to the axis of the applied tensile stress. Crack propagation under fatigue loading conditions (i.e., cyclic stress) is different from that of brittle fracture, described previously, which occurs under a constant load. This is so even though the crack progresses under plane-strain conditions with characteristic flat features. Of course, in real-life situations and components, stress applications are not so simple, and various types of loading may be superimposed and cracks can progress by several modes simultaneously, considerably shortening the lives of affected components.

Studies of propagating fatigue cracks show that stage 2 cracks often progress by an observable small increment during each cycle of tensile loading. In many materials, but not all, striations—parallel bands comprised of plateaus oriented normal to the stress axis and interconnecting ridges—can be observed at high magnifications running perpendicular to the direction of crack propagation along the crack path. Striations are characteristic evidence that the crack is of fatigue origin, although in very hard or very soft materials striations may not be visible.

While striations may occur during the peak tensile loading portion of the cycle, each striation does not necessarily represent one loading; otherwise fatigue life might be precisely predictable by measuring striation widths and calculating how many cycles would be required for failure. This is sometimes a useful indication but not always reliable. In some materials, striations are not visible. Striations are thought to correspond to plastic blunting at the crack tip followed by cleavage fracture and are, therefore, most apparent in ductile materials.

All parallel markings on fatigue fracture surfaces are not necessarily striations, however, particularly markings that are visible to the unaided eye. Some may be indentations caused by abrading of adjacent fracture surfaces during compressive segments of the cycle. Some microstructures have lamellar features (e.g., pearlite in low-alloy steels and eutectics) and may be misinterpreted as fatigue striations. During initial periods of crack propagation, striations are closely spaced but become more widely spaced as the crack progresses and unit stress levels increase. As the crack progresses and stress intensity increases, its surface texture becomes progressively coarser and has a more fibrous appearance.

Fatigue cracks in real-world structures and components do not always progress uniformly and predictably, as they do in laboratory-conducted tests. Each crack has a different "birthday" and will have unique characteristics. Depending upon loading conditions, operating circumstances, and the environment, crack progress may not be constant but may stop and go, slow down or speed up, with irregularly spaced intervals of growth. It is largely a random occurrence with progress based upon random influences and events. Consequently, striation counts become unreliable indications of elapsed time.

Fracture surfaces tend to reflect the history of the components and reactions to the environment (e.g., heat, moisture, corrosion, oxidation, periods of inactivity), much like growth rings of a tree. Such historical events and exposures leave visible traces of discoloration and other features. These are known as *beachmarks* or arrest lines and are observed visually as parallel markings on crack surfaces. They are often misinterpreted as microstriations which, when present, are discernible only under high magnification. Both, however, represent locations of the crack tip at different times. Figure 3.8 shows the appearance of a fatigue fracture of a steel bolt from a motorcycle exhaust manifold where bands of color tinting from progressive exposure to ambient oxidizing atmospheres at elevated temperatures produced characteristic beach markings.

Fatigue cracking usually initiates at numerous sites along the surface of a part. Because of stress level and many other conditions, not all progress to stage 2 propagation. But many of them do and, for a brief time into stage 2, independent crack fronts exist. At some point in stage 2, the several competing cracks usually link up to form a single crack front. When they do, the thin partitions of material separating them crack or fracture, resulting in a series of stepped plateaus, where plateaus represent crack surfaces and steps the partitioning ligaments. These "ratchet marks" are useful in identifying the fatigue mechanism and relate to multiple fatigue origins.

Existence of a subcritical fatigue crack in a part does not necessarily signal imminent failure. Some time may be required for it to grow to critical size. In addition, the loading conditions that initiated and propagated the crack to its present size may no longer exist, equipment

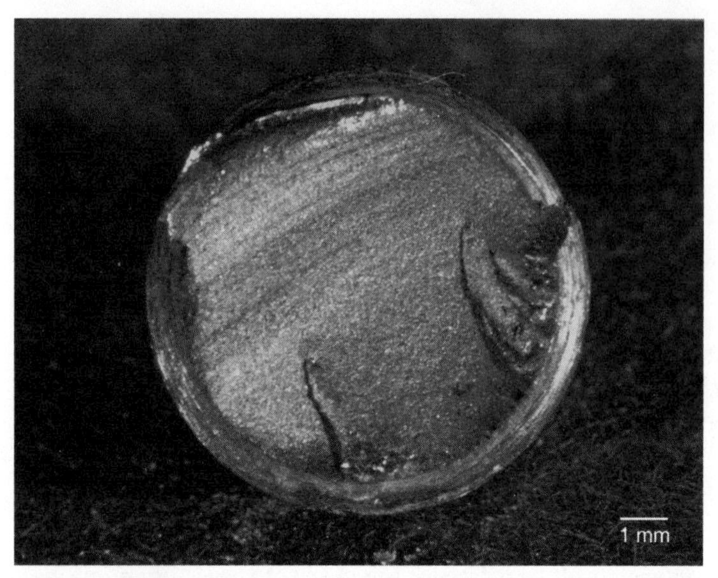

Figure 3.8 Fatigue fracture of steel bolt subject to intermittent exposure to elevated temperatures.

modifications may have shifted stresses to other members or locations, or perhaps environmental conditions have changed.

In the author's experience, a vibratory ball mill had operated continuously for some time without mechanical problems. During milling experiments evaluating the response of material requiring significantly longer residence within the mill chamber, its steel spring supports began failing by fatigue. Often, failure of replacements would occur within days. These unusual occurrences were traced to the use of cooling water sprayed upon the milling chamber by the operators to keep it from overheating during the prolonged experimental runs. When the cooling water was discontinued and integral cooling coils installed, spring failures ceased.

The existence of fatigue cracks in a component and their progression should be carefully monitored, however, to avoid catastrophic failure. This requires an understanding of the stresses being applied, the fracture properties of the material, information on environmental effects, and other factors.

Figure 3.9 shows the appearance of fatigue cracking in an aluminum fuel tank baffle that had been subjected to repetitive sloshing of fuel from vehicular motion. Here, the cracking initiated just above fillet welds joining the baffle member to the tank in a T configuration. Cracks progressed from both sides, as shown. Multiple origins, corresponding to weld bead irregularities along the interface of the weld beads and baffle, account for the jagged crack surface.

Figure 3.9 Fatigue crack in aluminum fuel tank along fillet weld joining internal baffle to tank bottom.

Figure 3.10 shows fatigue cracks in a cast aluminum alloy housing that had connected a diesel engine to an electric power generator. Once again, cracks originated from both sides of the member. These examples of actual fatigue cracks indicate typical complexities in crack appearance, texture, and multiple origins, and also show configurations of beachmarks and ratchet marks in some regions.

The third stage of fatigue cracking relates to the point at the end of the propagation stage when the growing crack has either attained critical size for the material and loading configuration or has reached the size where load-carrying capacity of the component is affected and it can no longer mechanically sustain the applied stress. At this point, failure usually occurs spontaneously through simple overload.

This final portion of the fracture will probably have a different appearance and will tend to be oriented in a 45° direction to the fatigue fracture. The final fracture will appear ductile or brittle, depending upon the material's properties, stress level and mode, environment, and other variables. Its size, shape, and location reveal information about the loading configuration and can be useful in analyzing the occurrence.

Fatigue life estimates. Fatigue life of a component is a function of stress level, stress state, the cyclic wave form, the metallurgical condition of the material, the environment, and other factors. As briefly noted above, fatigue crack development in real-life components is a statisti-

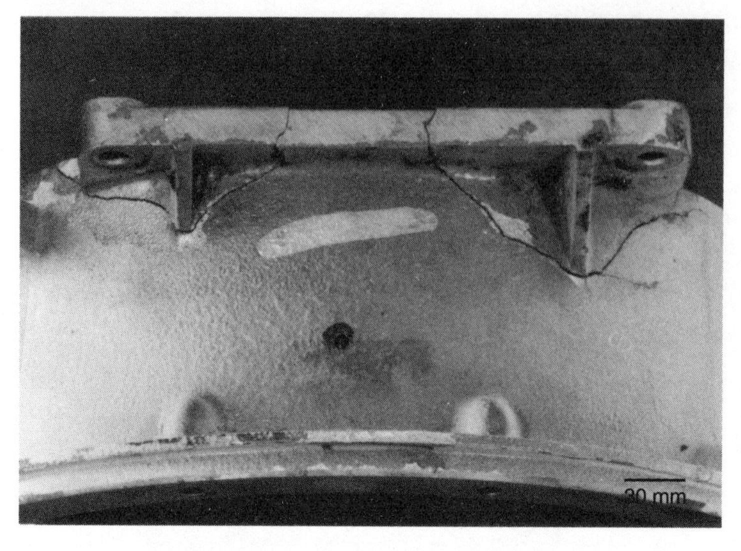

(a)

(b)

Figure 3.10 Fatigue cracking in cast aluminum motor-generator housing.

cally random process, as are its principal ingredients. Stress magnitude and cyclic profiles and their application are largely random as are the material characteristics and environmental conditions that contribute to crack initiation and its propagation.

Such a state of affairs does not readily lend itself to accurate predictions of initiation intervals, crack propagation rates, and projections of probable component life. Added to these difficulties is the tendency for small differences in component geometry and exposure conditions to have major effects upon fatigue life and behavior. This is even evident in laboratory-conducted tests under controlled conditions.

Nevertheless, fatigue testing is an essential activity, as it provides baseline information, insights into materials behavior under controlled conditions, and guidance in improving fatigue life. Many kinds of tests are conducted. Today there are two general types: those concerned with crack initiation and propagation to failure (the traditional types of fatigue test), and, more recently, those concerned with propagation characteristics. The latter use fracture mechanics techniques to determine crack-growth rates, proceed in a manner similar to those used for developing K_{Ic} data, and use similar precracked specimens.

In the more traditional tests conducted on uncracked, smooth-bar specimens, loading can be axial (simple tension + compression cycling), where the load varies from some maximum tensile level to maximum compression, or the load may be varied from maximum tensile to minimum tensile, or through bending. Tests are conducted on a series of identical specimens over a range of maximum tensile stress amplitudes, where the number of cycles to failure varies with the tensile load. Results of these tests (cycles to failure under given stress levels) are usually plotted as maximum stress amplitude versus cycles to failure (S/N curves) and have the general appearance shown in Fig. 3.11.

Different materials and alloys respond differently after a large number of cycles, as shown by the two curves. S/N curves for steels and titanium tend to flatten or become asymptotic to some lower stress level ("runout"), indicating virtually no limit on number of stress cycles below that stress level and suggesting infinite life. This leveling-off stress is referred to as the *fatigue limit* or *endurance limit* for a material. S/N data are usually depicted with N values in a logarithmic scale to more clearly show behavior at high stress levels, where fewer stress cycles are required for failure.[103]

Many variables influence the shape and location of these curves. The presence of notches in the specimen lowers the applied stress for achieving a given number of loading cycles. Also, the mean stress (one-half the algebraic sum of the maximum + minimum stress) has an effect. When a cyclic stress is superimposed upon a constant mean stress, the maximum value increases. Consequently, increasing the mean

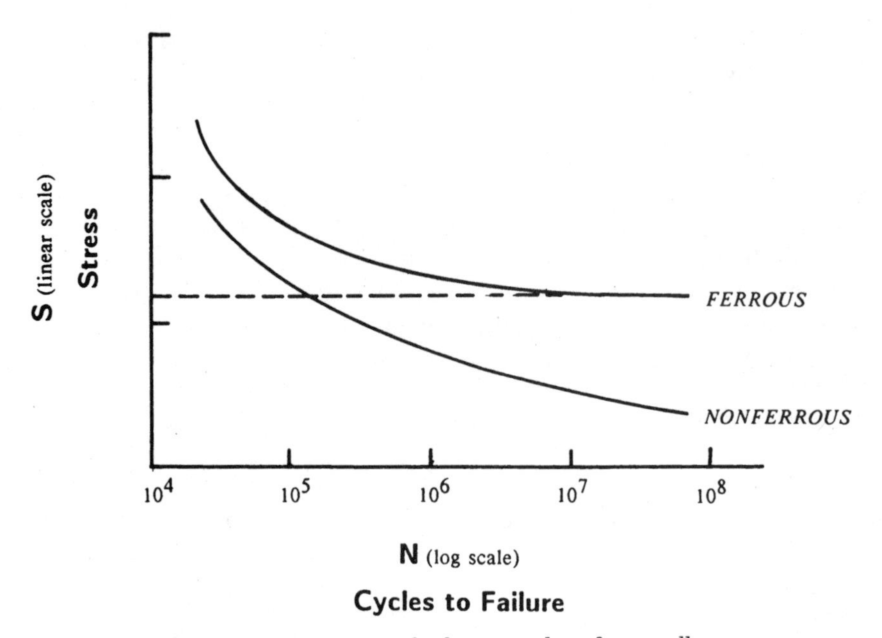

Figure 3.11 Typical S/N fatigue curves for ferrous and nonferrous alloys.

stress decreases fatigue life to less than that for zero mean stress (corresponding to complete stress reversal from maximum tensile to maximum compression). Test environments can have similar effects, where high humidity and other atmospheric contaminants, for example, tend to decrease fatigue life.

Most nonferrous materials do not show a definite endurance limit, as do steels. This is shown in the lower curve. For long life in these materials, stress must be continually decreased, making it necessary to define fatigue life in terms of stress and number of cycles to failure.

Tests for measuring crack-growth rate (da/dN) are more involved.[104] These determine rates at which subcritical cracks grow to critical size under given cyclic loading conditions. Specimens are precracked (by fatigue) and in the tests crack length is measured as a function of number of loading cycles. Results are expressed in terms of crack-growth rate per loading cycle (da/dN) and the fluctuating crack-tip stress-intensity factor (ΔK).

This is usually depicted graphically in log-log format, with ΔK representing $K_{max} - K_{min}$ for a constant R value (where $R = K_{min}/K_{max}$, the load ratio). See Fig. 3.12. Below some low value of K, the crack-growth rate becomes very small, and this stress-intensity value is denoted as the fatigue crack-growth threshold (ΔK_{th}). This value offers practical guidance in limiting maximum permissible loads for critical rotating components, for example, where crack extension during operation

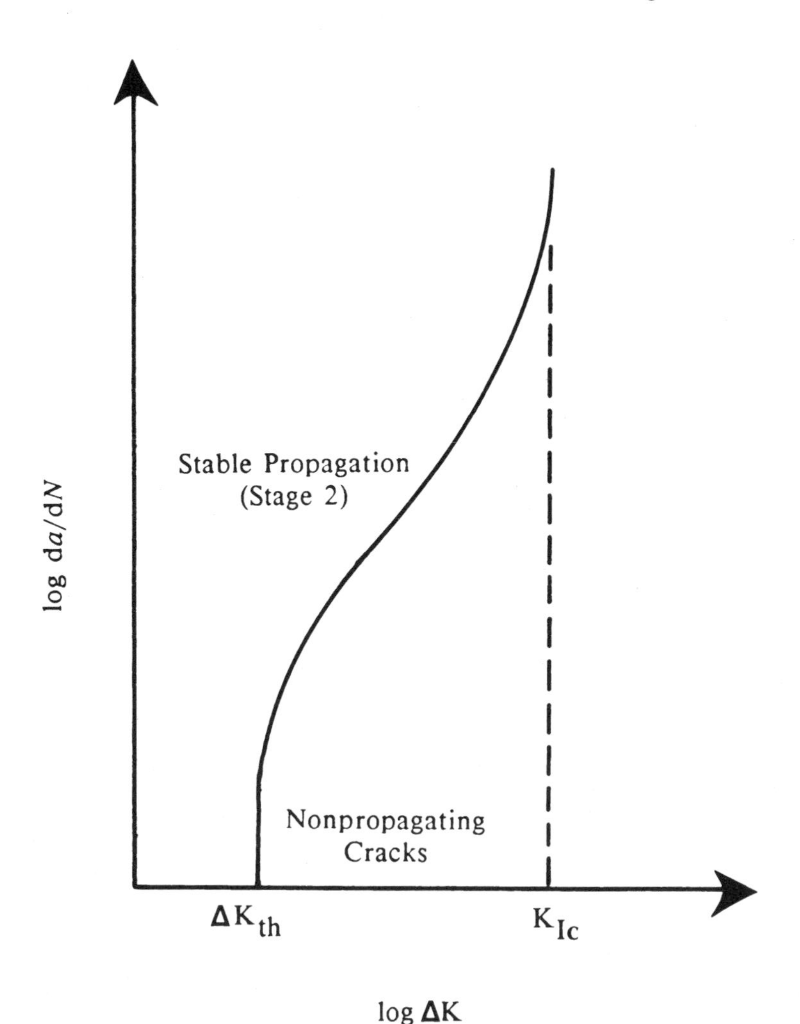

Figure 3.12 Crack-growth rate as a function of *K*.

would be particularly damaging. Similar crack-growth behavior is observed under stress-corrosion cracking (sustained loading) conditions, where the crack-growth rate (da/dt) is plotted as a function of the instantaneous stress-intensity factor K_I and where the threshold stress intensity becomes K_{ISCC}.

Fracture mechanics techniques for determining crack-growth characteristics provide a quantitative basis for estimating maximum tolerable flaw sizes in achieving desired design life. This opens questions of detectability of flaws of this size by available inspection methods and inspection accessibility requirements. Application of these techniques to safe-life and fail-safe design concepts is discussed in a later section.

High-temperature fatigue. Fatigue failures at elevated temperatures resemble those at lower temperatures.[105] As temperatures rise, a metal's mechanical strength decreases, and so also does its fatigue strength, as expected. Susceptibility to other attack or failure mechanisms, such as oxidation and hot corrosion, also increases as exposure to elevated temperatures increases, and these effects can drastically shorten fatigue life. Also, normally protective surface scales that form at elevated temperatures can spall off under fluctuating stresses and cause accelerated or breakaway high-temperature oxidation or corrosion.

Results of laboratory fatigue tests conducted at elevated temperatures show that creep probably does not play a major role under conditions of complete reversal cycling (maximum tensile to maximum compression, where mean stress = 0). For elevated temperature *service,* other failure mechanisms, such as stress-rupture, become dominant. At these temperatures, fatigue strength is affected more by total time under stress than by number of stress cycles, as metals continually deform under load at these temperatures where stress-rupture effects dominate. In fact, at elevated temperatures, stress-rupture strength (the stress that may be sustained by a metal at elevated temperatures for a given time without failing) is often lower than fatigue strength. Stress-rupture in many alloys is preceded by formation of internal voids that often interconnect to form a necklacelike defect. Fatigue cracks readily propagate from these sites.[106]

Thermal fatigue. Fatigue cracking that occurs from stresses generated as a result of heating and cooling cycles imposed upon the part is known as *thermal fatigue.* These are thermally induced stresses as distinct from fatigue at elevated temperatures, which is a result of cyclic strain.[107,108] However, there is a tendency to term all fatigue-related indications at elevated temperature as thermal fatigue or thermal shock.

Thermal fatigue, as rigorously defined, requires application of heat to a component under restraint. Localized heating generates a compressive force from the expanding metal which, upon cooling, contracts under restraint and induces tensile stress. Alternating heating and cooling cycles produce alternating tensile stresses that can lead to fatigue cracking (e.g., of turbine blades and similar components). This can be minimized through more uniform heating, elimination of sharp edges and regions of stress concentration, and use of expansion joints where other corrective measures are not feasible.

Corrosion fatigue. *Corrosion fatigue* is a description for cracking occurring under simultaneous action of corrosion and fatigue.[109] This situation is distinguished from stress-corrosion in that, while both are environmentally assisted crack-growth mechanisms, stress-corrosion occurs under a sustained static load, whereas corrosion fatigue occurs under the combined action of a cyclic load and an aggressive environment.

Fatigue life can be drastically reduced under such combined conditions. In a sense, though, practically all fatigue failures occur by corrosion-fatigue, as environment almost always plays a role in the process. This is demonstrated in fatigue tests conducted in a vacuum where greater fatigue life is observed than in ordinary environments and ambient conditions.

The combined effect of simultaneously operating failure mechanisms is probably nowhere more dramatic than in fatigue tests conducted in gaseous hydrogen environments. Losses in fatigue life of 90% have been reported in moderate-strength pipeline steels in the presence of notches or other stress-concentrating geometries. Crack-growth rates of more than 10 times also have been observed and in some cases may be over 100 times greater than that in a reference environment. Hydrogen effects on fatigue properties are reported to be affected by many factors including stress concentration, stress ratio (R), cycle frequency, hydrogen partial pressure, test temperature, and presence of other gases.[110] It is somewhat reassuring, however, that crack-growth acceleration by hydrogen was completely inhibited by 100 ppm oxygen and nearly completely by 2% carbon monoxide or 2% sulfur dioxide. These are reported to apparently block hydrogen access to the steel by preferential absorption.[111]

These brief descriptions will serve to acquaint the reader with the nature and magnitude of problems that can result from some of the more common, and more devastating, failure mechanisms. Precautions and procedures for minimizing their effects have been noted in passing, and the cited references can provide details that are beyond the scope of this book but that are necessary for a more complete working knowledge of these and related failure mechanisms. Additional information on some of the more applicable failure avoidance strategies for dealing with these problems and others is included in later sections of this chapter and in the next.

In the meantime, however, a recurrent theme frequently associated with mechanical failures deserves attention. This involves the processes and techniques commonly used in joining metals and alloys and the resulting welded joints or weld deposits. Of particular relevance to our discussions are why they are implicated in so many failures and why they are frequent sites for fracture initiation.

Welding and related issues

Vulnerable sites. Although welding has been an accepted method for joining metals for well over a half century, too many engineers are still inadequately informed in its fundamentals. For laypersons, the level of knowledge is significantly lower. Despite its wide use and acceptance and the average person's general familiarity with the concept, basic welding principles are not well understood. Welding is often the site of fractures and other problems; therefore, today's engineers, managers, and even nonprofessionals employed by industry cannot afford to be uninformed about a field that so pervades our technological world.

Qualified and highly competent engineers have spent entire careers in the welding field, many in welding research and development, who will unhesitatingly admit their lack of understanding of many aspects. So we must not let superficial familiarity lead us to believe that that's all there is to it. If we are to develop the capability for avoiding failures, particularly of the serious kinds we have been discussing, we must become more knowledgeable in the intricacies of these processes and materials.

We cannot avoid the fact that there have been serious failures that in one way or another are associated with welds. Published case histories are full of them. Some may be the result of incorrect diagnoses, but many failures involve welds. Considering that it is the principal means for joining metals, this probably should not be surprising. After components to be welded have been aligned, clamped into position, and the welding run completed, the finished joint is usually considered equivalent to original components in properties and characteristics. For all practical purposes, this may be an acceptable viewpoint for most common applications. But there *are* differences in and around that welded joint which, for some applications, loading configurations, and environments, can determine whether the welded part survives.

Welding processes join materials that usually have been carefully produced under controlled conditions and according to industry consensus standards or government agency specifications. These dictate characteristics of the metal form, such as dimensions, acceptable limits and permissible ranges for chemical composition, minimum properties, identifying marks, and capabilities for meeting various service requirements. Examples of these requirements are ductility, notch toughness, corrosion resistance, and other acceptance and performance criteria. Welds joining these materials are expected to match the properties and characteristics of the plate, tubing, structural shape, or casting.

When it comes to welding these materials—a process that can be even more critical to the life of the component being fabricated—we are often not concerned about how it is done. Too frequently, inadequate attention is given to whether the weld metal that will be subsequently regarded as possessing all the qualities of the original components will, in fact, have those qualities. Will the composition and microstructure be appropriate? Will the properties be adequate? Will the welded joint be free of harmful defects and suitable for the intended service?

Certainly, for critical applications, code-qualified welders must do the work, pedigree documentation must exist for all materials, quality-assurance practices must be followed, timely inspections made by certified inspectors, and each step must be checked and rechecked. But, for the most part, these are exceptions.

Earlier in this chapter we described several principal damage mechanisms. These were chosen for several reasons. They account for most failures of engineered structures, components, and systems. Besides this, their fracture mechanisms and failure modes are insidious in that they involve cracking that occurs with little or no outward evidence that it is in progress. Conditions required to trigger the cracking are normal everyday operating conditions, and it occurs under stress levels that are typical of usual operating loads for their applications.

It is unfortunate that welding creates conditions that are particularly susceptible to these failure mechanisms. We cannot, of course, in the limited space available here, attempt to address even a few of the issues involved in producing high-quality failure-resistant welds in the wide range of metals and alloys in use and by the array of welding processes currently available. This is well beyond our intended scope. Our intention here, however, is to consider some of the conditions that make welds especially vulnerable to the fracture mechanisms described. Since these represent mechanisms that are responsible for or are at least implicated in most failures and since welds are probably the most vulnerable location for their occurrence, any improvement in understanding of why welds are prime targets can be of significant benefit in avoiding failures.

Welds have characteristics that are different from other metal and alloy forms used in engineering structures and components. It is these differences and peculiarities that lie at the root of most, if not all, of the problems and susceptibilities that make welds a primary site for various fracture mechanisms. Six categories of such features have been identified as predominant. Undoubtedly there are others, but these account for most of the problems. Keep in mind that not all welds, metals, and alloys are affected by all of these characteristics:

1. Geometric effects
2. Discontinuities
3. Compositional and microstructural heterogeneity
4. Residual stress
5. Thermal effects
6. Hydrogen inoculation

Geometric effects. The purpose of welded joints is to transfer loads from one member of the structure or component to another to provide mechanical and metallurgical continuity. Welds are connectors or links, where surfaces, parts, and components intersect. To designers, such intersections are logical and economical locations for welds. However, such locations often are the highest stressed in the component. Therefore, these "natural" locations for welds, regardless of their quality and integrity, can make them vulnerable and preferred sites for failure.

Welds often become sites for accommodating and absorbing dimensional and distortional corrections. Out-of-round cylinders, warped structurals, incorrectly toleranced, or cut members are often welded, then pulled, pushed, or beat into shape or the desired configuration after welding is completed. Frequently this is accompanied by localized heating to facilitate the corrective forming. This largely unscheduled metal processing can result in severe plastic deformation in and along welded joints, which further complicates an already complex residual stress situation and which further affects microstructure and mechanical properties.

Fillet welds and overlapped joints are common types that are easy to make, require only minimal welding skill and joint preparation, can tolerate high levels of welding power for high weld deposition rates (a productivity benefit) without burnthrough, and are low in cost. Their fatigue properties are poor, however. They concentrate applied stresses at their weakest regions, the softened heat-affected zones adjacent to the weld bead. In addition, they are partial-penetration welds with sharp builtin notches at the weld root, a prime location for initiating cracks.

There are alternatives to these fatigue-prone joint configurations but they cost more to make in terms of joint preparation, joint alignment, welding skill, and production rate. Butted joints are far superior in fatigue strength. Their use, plus repositioning the joint away from highly stressed corner locations, can be a major step in avoiding weld-related fatigue failures.

Fatigue life is not the only consideration, as the sites described constitute notches and, depending upon fracture properties of the material, service temperature, and other factors, can become sites for brittle fracture. These highly stressed locations are also preferred sites for stress-corrosion (in susceptible materials and environments) and for other forms of corrosion and hydrogen embrittlement.

Weld beads deposited by most commercial welding processes have naturally rippled surfaces, and these ripples and ridges are notches and stress concentrators. Attempts to improve fatigue life by removing these irregularities can backfire, however, if the method used for removal merely introduces other surface defects, such as grinding marks, scratches, and gouges. Attempts to smooth bead surfaces can also thin the weld and decrease its strength and bring subsurface flaws to the surface where they can be detrimental to fatigue life.

Discontinuities. Welds are subject to many kinds of defects. At least ten categories have been identified. Some are introduced at the weldment design stage, some are associated with the welding operation itself, a characteristic of the metal being welded or filler metal being introduced, and some are related to residual stress. The defects may be located within the weld itself, the heat-affected zone of the weld, or both. Some may be caused by the weld but are manifested well outside the immediate weld zone, as in strain-age cracking or cracking as a result of hydrogen embrittlement.

The ten categories of weld defects are as follows:

1. Incomplete penetration
2. Inclusions
3. Undercutting
4. Liquation (hot) cracking
5. Delayed (cold) cracking
6. Stress-relief cracking
7. Strain-age cracking
8. Porosity
9. Embrittlement
10. Structural notches

Some defects can be controlled or minimized by the welder or welding machine operator; others are characteristics of the material. For example, the first three on the list may appear to reflect welder competence; however, these problems are often traceable to design deficiencies. Insufficient plate bevel angle or root gap, or positioning restriction due to inadequate joint access, can cause these conditions, as can insufficient welder skill. Others reflect inadequate procedural control.

Some of the listed defects are potential problems of themselves (4 through 7). These are cracks that can weaken the component or can grow under operating conditions and loading and lead to structural failure through brittle fracture or fatigue. The others are detrimental in setting the stage for initiation of fatigue cracks or other cracking mechanisms.

Some metals and alloys are more tolerant of the welding process than others. This includes weld metal as well as base metal. Usually, the more complex the alloy in terms of strength level and other performance aspects, the less "weldable" it is. Weldability can be considered to vary inversely with a material's strength and performance. There are many reasons for this. The point to remember is that welding high-performance alloys for any application, but particularly for critical applications (as high-performance alloys usually are used), must be approached with caution. Chances are that it will not be straightforward and will be prone to fail. This does not imply that such welds are impossible, cannot be successfully made, or should not be attempted. It only says that they have little tolerance for error and require understanding of all the factors involved, plus good judgment and practical welding experience.

Compositional and microstructural heterogeneity. Fusion welding processes employ concentrated heat sources, usually shielded electric arcs. Their purpose is to locally establish a melted region at the edges of the parts to be joined and, frequently, to melt the filler metal used to bridge the gap and constitute the weld nugget. The weld "puddle" formed in these processes is, in effect, a miniature crucible.

In this "crucible," complex chemical and metallurgical reactions take place; many of them parallel and duplicate those of steelmaking or alloy production mill processes. However, in the weld pool, reaction time is necessarily brief as the welding heat source progresses along the joint leaving previously reacted products behind to solidify and cool as it continues these reactions in new adjacent locations. The crucible and mold here is the metal being welded, which becomes part of the joint itself.

From time to time, welds have been referred to as miniature castings, but this is not an altogether correct analogy. Both involve metal melting and resolidification, but welds solidify epitaxially from the base metal which becomes an integral part of the completed joint, whereas, in castings, the mold does not become part of the final product.

In the welding process, some of the base metal becomes melted into the weld, along with whatever supplemental filler metal is added. The two mix very briefly, then are forced to solidify and cool as the heat source progresses down the line. In view of this scenario, it is not surprising that weld deposits are sometimes flawed. What is most surprising is that so many are not.

Laterally, across the welded joint, metal temperatures range from ambient to well above the melting point. There has been a largely random intermixing of not only base material and filler metal, but elemental additions in the form of metal powders may have been introduced as supplemental alloying constituents and also mixed in. Fluxing ingredients in some processes have also entered the weld pool, melted, reacted with both metallic and nonmetallic constituents and the atmosphere and have formed new substances and compounds that escape to the atmosphere or float to the top of the bead as slag to protect it from oxidation. To further complicate matters, additional weld beads are often superimposed on top of previously deposited ones. Portions of these previous beads are reheated and remelted into the new deposits and become part of it.

All of this produces anisotropic material that is unlike any other. Yet, it is considered an integral structural component having definable, measurable, and even predictable properties. It is trusted to hold the welded components together as it is now part of the original tubing, plate, sheet, structural member, or casting itself.

However, its composition is not the same, especially on a microscale, where incomplete mixing has resulted in alloy-rich and alloy-lean regions and segregation of constituents.[112] These heterogeneities produce microstructural differences from place to place, such that the microstructure of the base metal is not duplicated throughout the weld that joins it. Weld metal chemical analysis performed on weld metal samples reveals average compositions and does not reveal the localized (microscale) regions of elemental segregation.

Since weld metal composition and microstructure are not uniform throughout, there is a good possibility that some regions will have properties different from those of other regions, as well as differences in susceptibility to fracture, corrosion, and other forms of metal degradation. Weld heterogeneity is responsible for segregation of low-melting-point interstitials, impurities, contaminants, and trace elements that form liquating films. Figure 3.1 is an example. These can cause hot (or liquation) cracking in welds and in partially melted regions of heat-affected zones as shrinkage stresses from solidification of higher-melting-point constituents are imposed upon still-molten grain boundaries, tearing them apart. Some types of weld cracking do not occur in the deposit while it is being made, but in deposits reheated by subsequent weld beads placed upon it. This is a result of combined action of the reheat cycle and solidification stress. Microfissures in austenitic (3XX series) stainless steels are frequently of this type.

These anisotropic microcomposite materials (welds) respond differently than wrought base materials under exposure to elevated temperatures, and such applications require special consideration. Behavior under creep and stress-rupture, particularly ductility, can be inferior to that of wrought forms of similar average composition. These differences must be offset through placing welds in cooler regions or using filler metals of modified, perhaps enriched, compositions, or simply designing to accommodate weld properties.

When weld homogeneity is important to the life of the part, post-weld heat treatments often are employed. These afford time for alloying constituents to interdiffuse and for stresses and microstructures to equalize. It is not frequently done, as it is costly and can produce distortion as residual weld stresses equalize throughout the part.

Weld heterogeneity is largely a natural consequence of solidification, the compositions involved, and the heating and cooling rates associated with the welding process. Therefore, it is not controllable by the welder or welding operator. The weld designer also has little control over it. However, an awareness of its existence and attendant susceptibilities under various circumstances could help to forestall problems in specific applications that might stem from it. For example, nonfusion type welding processes (e.g., diffusion bonding) or lower-temperature joining methods (brazing) might be used. The welds themselves may be relocated to outside the environments or conditions of concern, or solutioning heat treatments may be applied to homogenize the weld deposit.

Residual stress. This is a consequence of localized heating, cooling, and solidification of welds and related plastic strains. Contraction that takes place after welding is proportional to the thermal gradient and volume of weld metal solidifying and will differ from one direction to the other. It is important when evaluating distortion and other effects to realize that these contractions operate in three dimensions, meaning that resulting stresses are triaxial. This can influence material response under service loads and other stresses.

Residual stress as a consequence of welding can be of yield-point magnitude and plays a major role in failures of the types discussed earlier. Tensile stress is required for initiation and propagation of stress-corrosion cracks, hydrogen-induced cracking, propagation of brittle fractures, and for initiation and propagation of fatigue cracks. Residual stress in the vicinity of welds prone to failure because of adverse geometric configurations is especially damaging. Figure 3.9 is but one example.

It should be mentioned that while we have been referring in this section to *welds,* the same comments also apply to metals deposited by welding processes for purposes other than joining metals. Overlaying, hardfacing, buildup of worn parts, and similar operations are examples. In addition to considerations of metallurgical and microstructural compatibility of such deposits with the base metal, the effects of both residual stress and stress from differential expansion and contraction of dissimilar metals bonded to one another must be considered.

Residual stress also contributes to the additional difficulty of welding heavy-section material. Increased rigidity offers limited opportunity for plastic deformation to relieve builtup stress. Other factors are also involved in heavy-section welding. Additional thermal mass will tend to make cooling rates and solidification rates higher, and the increased restraint of heavy sections can produce plane-strain conditions in and around microfissures, cracks, and flaws and promote crack growth as a result of increased resistance to plastic deformation at the crack tip. Hydrogen embrittlement and associated delayed cracking in susceptible materials (high-strength steels, for example) can be aggravated in thick sections.

Thermal effects. Some of these have already been mentioned. Of principal concern are effects of heat cycles imposed upon the materials (base metal and weld metal) during welding. This involves rates of heating and cooling, time at peak temperature, and numbers of such cycles. These factors are especially relevant in materials like low-alloy steels in which mechanical properties depend upon thermal treatments. Welding can excessively temper some microstructures and lower strength, or rapid quenching in heat-affected regions (by thermal mass of the component) can produce brittle microstructures having poor toughness and inferior crack propagation properties. These possibilities must be taken into account in considering welding such materials, and procedures have been developed for this.

Throughout the welding process, components being welded are exposed to elevated temperatures. While the immediate weld pool is protected from atmospheric contamination by flux, slag, or shielding gas, surrounding regions experiencing high temperatures may be subject to oxidation and other atmospheric contamination. This is a special concern along the underside of welds (often inaccessible for protection) and surfaces of previously deposited beads in multipass joints.

Damaging oxide inclusions can be produced in welds intended for elevated temperature service (jet engine parts, for example) where previously deposited weld metal is not adequately cleaned of high-temperature oxides and films before welding over them. Many so-called superalloys contain chromium, aluminum, titanium, and other strengthening elements that form tenacious refractory oxides. Often, an effective solution to eliminating this problem is to weld within protective sealed enclosures filled with shielding gas or within vacuum chambers using suitable processes, such as electron-beam welding. Where this is not feasible, supplemental extension (leading and trailing) shields can be used to extend the protective gas cover over vulnerable heated regions.

Thermal treatments can have a beneficial effect in welding. Preheating and post-heating are useful in minimizing residual stresses and in avoiding cracking during welding of brittle materials and heavy sections. They are also used in achieving the necessary weld metal and heat-affected zone cooling rates for maintaining desired weld properties through microstructural control. And preheating is an effective means for removing moisture on materials to be welded. The importance of this is discussed in the next section.

Hydrogen inoculation. Welding operations offer a prime opportunity for inoculating welds with damaging hydrogen. Moisture is the principal source, and it can come from the metals being welded, water vapor in the shielding gas, atmospheric humidity, welding equipment coolant water leaks, or from filler materials. Organic contaminants such as oils and cutting fluids are other sources, and welding over moisture-containing corrosion products such as rust can inoculate with hydrogen. Most fluxes are hygroscopic and absorb moisture from the atmosphere and can be another source. Some flux ingredients such as cellulose, used in some types of steel flux-covered welding electrodes (*stick* electrodes), and various organic constituents and binders decompose in welding arcs, again introducing hydrogen.

As in the example of welding maraging steels described earlier, hydrogen may also be introduced directly into the welding pool through the filler metal. Hydrogen-inoculating processes such as electroplating, pickling, and cathodic cleaning are common causes. These can be particularly damaging in steels of moderate to high strength levels. The best precaution for avoiding these problems is to eliminate processing operations that introduce hydrogen, but this is often impracticable due to wire drawing difficulties often encountered in high-performance alloys. Extended baking treatments of the finished wire can offer relief but may not effectively eliminate all the hydrogen.

It should be mentioned in this context that much of the mild steel weld filler materials marketed today are furnished with a copper-plated surface to protect from rusting. This is usually not a problem, as steels of ordinary strength levels (i.e., plain carbon or "mild" steels) are not susceptible to hydrogen embrittlement, as explained earlier.

Welding arcs decompose organic compounds and other substances, including moisture, with atomic hydrogen a by-product. Hydrogen solubility in molten metals is substantially greater than in solidified metals, and weld metals can act as sponges for absorbing hydrogen produced during welding. The consequences can be serious, as discussed in an earlier section. The major problem is delayed cracking in steels of susceptible hardness and strength levels. In softer alloys and those not susceptible to delayed cracking, entrapped hydrogen can produce porosity and general unsoundness. Scattered small pores in welds are usually not detrimental, but when clustered, aligned, or large in size they can degrade mechanical properties or become sites for crack initiation.

Weld filler materials intended for use on alloys susceptible to hydrogen embrittlement should be well protected from moisture pickup during storage and especially after opening the containers. Flux-covered electrodes, flux-cored wires, and submerged-arc fluxes should be kept in heated ovens until ready for use to avoid moisture absorption and hydrogen embrittlement of the welds. This is recommended for welding of all metals and alloys, not only those susceptible to hydrogen-induced cracking, as moisture pickup can cause many welding problems.

Welding of susceptible steels should use only "low-hydrogen" grades of filler materials.

Welding information sources. These brief comments cannot do more than instill within the mind of the reader an awareness of some of the reasons why welds have been and can be sites for fractures that have led to failures of all kinds and degrees of severity. Each aspect can occupy an entire volume and not exhaust the subject, or even cover it comprehensively.

There are many sources for further information, and some are cited in the references.[113–122] Publications and activities of the American Welding Society (Miami, Florida), The Welding Research Council (New York, New York), The Welding Institute (Cambridge, England) and its U.S. affiliate, the Edison Welding Institute (Columbus, Ohio), are primary sources for welding information. Publications and activities sponsored by other engineering and metallurgical societies are also sources and include ASM International (ASM denotes the American Society for Metals), The Metallurgical Society of AIME (American Institute of Mining, Metallurgical, and Petroleum Engineers), The American Society for Testing and Materials, and The American Society of Mechanical Engineers. Engineers and managers involved in any way with metals joining operations should become knowledgeable of the information and benefits provided by these organizations.

Life-Cycle Management

Rationale and concept

The problem of mechanical failures, as well as failures of all engineered systems and components, requires attack on many fronts. Failures are unwanted, unexpected, and unintended events—the exceptions. All engineered systems, of any size or complexity, are designed, constructed, or manufactured to function satisfactorily and to be fit for their intended purposes, that is, without failing. For the most part, engineering practices usually work well and result in trouble-free, safe, and reliable products and systems that are fit for their intended purposes and fulfill their functions without failing.

But things do not always work out this way. Products and systems sometimes do not perform as designed or intended. They can be misused and abused, used in a way unforeseen by designers, used beyond their useful lifetimes, or pushed beyond their limits. Environmental effects can be more aggressive than anticipated; combinations of effects and use functions can lead to unexpectedly rapid deterioration. And defective connections, assembly errors, missing parts, and material deficiencies can degrade performance and cause failures. Although these are the exceptions, they are the focus of our attention here, as our goal is to reduce their number and severity.

The sudden and unexpected character of failures stems from the complex causation networks behind them. Other contributors are the frequently unsuspected susceptibilities of engineering materials to deterioration and damage and our yet-incomplete understanding of them. For many years, and in recognition of these uncertainties, engineers have designed and constructed machines and components following conservative practices.

Traditionally, these have involved assumptions of worst-case mechanical properties and response to service environments, applications of generous safety margins, and proof tests. Completed prototypes are subjected to tests under anticipated service conditions, and once into production samples are inspected to obtain statistical estimates of some quality characteristic or performance expectation.

These traditional practices get into trouble when worst-case assumptions turn out to be overly optimistic, when safety margins do not take all risks into consideration, and when results of proof tests are not indicative of response under repetitive loading (as they never are). Quality control inspections may not check the most critical variables. The fact that these practices work as often as they do also generate in their users an unfounded sense of confidence. Traditional engineering approaches also make extensive use of handbook data. Design loads carried by components are calculated from these values, whether it is a structural member operating at ambient conditions or a superheater tube. Actual components experience complex multiaxial and dynamically applied loads, whereas handbook values are derived from simply loaded laboratory specimens. Corrosion-response tables based upon simple immersion experiments conducted with small specimens may not indicate the drastically higher corrosion rates experienced by the same materials when velocities increase or aeration is present.

Published values for fatigue strength probably were developed on laboratory specimens under sinusoidal loading, whereas actual stress cycles usually are far more complex. Handbook tables do not contemplate the array of possible dissimilar-metal combinations that are possible in real-life assembly applications and the range of seriously degrading effects that can result (hydrogen embrittlement and stress-corrosion, for example). These can occur under normal operating conditions and at tensile loads well below the handbook yield strength. These gaps in information are usually compensated for by safety factors and use of other conservative materials selection and design practices. However, damaging failures still occur. Obviously, their consistent avoidance requires more diligent attention.

In Chap. 2, we discussed the usefulness of failure analysis in forestalling engineering failures and recurrences. One of the reasons that it is such a valuable tool is because failures represent actual experience and response of the product or system to real-world conditions. No laboratory test can truly duplicate them. It is evident that we cannot rely upon failures for detecting weak points and deficiencies in design, although unscrupulous manufacturers have been known to do this. But recent developments in products-liability law have sharply curtailed this practice. Besides, even in the absence of liability threats, failures of most engineered products and systems can be economically devastating for other reasons, such as costs of unscheduled downtime, equipment damage, and loss of profits, sales, and customer confidence. So, failures teach valuable lessons and provide facts and information that are not usually available elsewhere. We must do all we can to avoid them; but when they occur, we cannot afford to ignore what they have to say.

Earlier in this chapter, we also described analytical procedures for evaluating hazards and vulnerabilities in manufacturing or processing operations, equipment, components, or entire engineered systems. These procedures help identify critical items that can cause or contribute to unwanted events, accidents, and failures. As useful as all of these analytical tools, procedures, and results of failure analysis are in avoiding failures, they are not enough. The web of failure causation is extremely wide and encompassing. We cannot wait for failures to occur to learn of susceptibilities so we can make design modifications or use better material.

This is especially true for large and expensive systems, such as power-generating stations, bridges, nuclear submarines, offshore oil drilling platforms, advanced weapons, petrochemical plants, oil refineries, cross-country gas pipelines, oil tankers, and commercial aircraft. In today's world, we cannot let such complex systems continue to operate as designed and constructed and simply "wait and see" how they fare, and then incorporate into new construction and replacements what was learned through forced outages, crashes, and disasters.

Many of the world's electric power-generating stations, petrochemical plants, and other industrial processing and manufacturing facilities have been in operation for considerably longer than their originally intended design lives of 30 to 40 years. All indications are that the world economic situation, tightening environmental regulations, and other factors discouraging substantial capital investment in new plants will continue throughout this century and perhaps well into the next. This will require continued operation of existing facilities for an even longer time, if it is possible to do so.

As pointed out for the case of fossil-fueled power plants, the purpose of life-extension activities is not to continue a plant's operation beyond its useful life, as this would result in inefficiency and reduced availability and even failures. But their purpose is to ensure full utilization up to the end of a plant's useful life.[123] Similarly, our broader interest lies in maintaining efficient use of products, plants, facilities, components, equipment, and other engineered systems and in forestalling failure, throughout their useful life.

The terms *design life* or *useful life* are no longer definitive, as periodic overhauls, selective replacements of components with improved versions, performance monitoring, and operational evaluations can significantly extend useful lifetimes. It is important, in discussing these matters, to understand that these extension measures are not emergency expedients such as are commonly found in Third World countries where, for economic reasons and sheer desperation, obsolete industrial plants are forced to continue operation well beyond their useful lives. Through technology advancements, the effectiveness and availability of life-extended plants may even exceed those of original installations. And it has been cost-effective to do so.

Although electric power-generating plants are often singled out as examples of applications of life-extending techniques—probably because of their size, complexity, critical need, and cost—there are many other examples, some only now emerging. They include commercial and military aircraft, bridges, ships, manufacturing and metalworking machinery, and petrochemical plants. Some common characteristics among these applications offer guidelines for applying these techniques to other fields. Since considerable expense is involved, benefits must be weighed against costs. Such decisions must also take into account the consequences of not doing so, or the costs and effectiveness of alternative action.

Hazards analysis will reveal susceptibilities inherent in the operation, product, or equipment and will indicate their severity and level of criticality. If a significant hazard exists, the operation, activity, or component associated with it and the reasons why should be identified through FTA and FMEA techniques. Often, these techniques will show how the hazard can be eliminated or diminished. If so, corrective action can be taken. If, however, the hazard, fault, or susceptibility to failure cannot be eliminated or substantially minimized, other avoidance measures must be considered. This is where the decision model of Fig. 3.2 can be applied.

In using fault-tree or event-tree analytical techniques, there is a tendency for the user to concentrate upon eliminating the "top" or upper events. It may be a natural inclination to focus upon the unwanted occurrence, but its causes usually lie several layers away. Key "ingredients" for the occurrence—and the most effective failure avoidance measures—are found at lower levels of the diagram. Failure avoidance should work its way up from the bottom of the fault tree.

Since we are confining our attention in this section to operating plants and equipment that are in service, as opposed to those in conceptual or design stages, our consideration will focus upon measures for keeping those plants and equipment operating without incident, that is, to ensure uninterrupted operation without unscheduled downtime or stoppages due to mechanical breakdowns. Descriptions for implementing the decision model of Fig. 3.2 have been purposely generalized and necessarily vague to accommodate a broad spectrum of applications. Each decision process will be unique as will its objectives, requirements, inputs, evaluation criteria, and results. Details must be developed by the user, but the tools and decision methodologies will work for a wide range of needs and applications.

Similarly, at this early stage of their development, some of the life-extension procedures derived and used in one industry may not appear to be directly transferable to another. However, most of these techniques can be adapted to other applications where incentives for avoiding failures and shutdowns are great enough, and the consequences of failures serious enough, to justify whatever effort is necessary to learn to apply them whenever needed.

It must be realized that life extension is a developing technology and is probably only now in its initial stage. It has evolved primarily in response to economic constraints that make replacement of existing power-generating stations undesirable and impracticable. A second area of active development is aging aircraft. Others are under consideration generally for the same reasons.

Failure avoidance is but one of the reasons for developing life-extension practices. Continued productivity, operational availability, and cost-effectiveness are other reasons. While strategies and methods used for specific industries and applications may be most relevant to them, most can be applied, with appropriate modification, to other industries and facilities.

Life-extension approach

Application of life-extension methodologies to an existing plant or system is not practicable without an organized plan, and this requires a series of critically important decisions. The inherent complexity of manufacturing and processing systems and possible failure modes and causation elements preclude any serious attempt on the basis of intuitive reckoning, no matter how simple it all appears to be. Some kind of formal decision analysis should be used; method and format are relatively unimportant. There never are sufficient funds to attend to everything that needs attention, and one of the first decisions that must be made is how to allocate available resources to a realistic program of failure avoidance. Even before that, a decision must be made to determine if such a program is warranted or needed. The effort will be doomed from the start if there are no compelling reasons for it.

A first task should be to evaluate the entire plant as a system. Why does it or should it exist? Perform a decision analysis on it using an approach like the one of Fig. 3.2, for example, with inputs modified to suit the situation. With this done, threats to its continued existence should be evaluated. This step should include *all* threats and hazards, whether involving material failures or not. It might include such possibilities as loss of market share, insufficient capitalization, labor unrest and personnel problems, increased bureaucratic interference, environmental compliance, adverse court decisions or impending litigation, embezzlement, natural disasters, and shareholder disputes. These kinds of threats and potential threats must be evaluated along with physical hazards to the business, its plant, and facilities.

A hazards analysis format (spreadsheet) may be an appropriate starting point, but a fault tree might provide a clearer perspective for study purposes. Keep in mind that these are not ends in themselves but merely tools and decision aids. Decision models can assist the decision maker to think correctly and about the right things, but only if they are thorough and conscientiously used. Their value is only as good as the effort that went into them. It is important that input for these decision aids is based upon realistic assumptions and existing conditions and situations.

With principal "faults" (unwanted events or conditions) identified, it should be possible to determine the most critical aspects of the operation or business. Some threats will be worse than others, some more likely than others. Risk analysis techniques use quantification methodologies and probability assessments for accomplishing this and are useful where numerical results are needed or desired. Quantitative methods are also applicable to fault-tree analyses and might be applied. We are dealing here, however, with basic concepts, and our primary goal is to create in the reader an awareness of the need for and value of failure avoidance practices. Detailed analytical steps that might be required for thorough assessments of specific situations are beyond the scope of what we can hope to accomplish within these few pages.

Presumably, hazards or threat assessments will result in identification of several possibilities, and these should be rated as to criticality and likelihood. Obviously, the most likely *and* most critical items require primary attention. Since our emphasis here is upon mechanical failures, it is possible that the analysis will identify one or more events or conditions of this nature. Many factors are involved in identifying these and should include an evaluation of the probable consequences resulting from the occurrence of such events or conditions. These probably would have been considered during the initial criticality assessment, but here the objective is not only to identify the most critical or damaging event but also to determine what is required to forestall it.

For example, the most critical potential problem that may be identified is possible injury to neighboring residents from rupture of a tank used for storing toxic chemicals, located on adjacent plant property. One approach might be (1) frequent inspection, (2) monitoring of the structural integrity of the tank, (3) equipping it with state-of-the-art sensors to detect the onset of problems, (4) erecting spill barriers, and (5) developing evacuation procedures. A more effective solution might be found in reassessing the need for maintaining large inventories of the toxic chemical and possibly eliminating it altogether, or in considering process modifications to use less of the chemical or not require interim storage at all.[124–126]

Where there is significant risk of personal injury or damage to the surrounding community, it is almost always preferable to choose the "inherently safe" route and eliminate the hazard or at least significantly diminish it. This is in contrast to the "extrinsically safe" approach where failure avoidance depends upon monitoring the hazard or relying upon sensing devices, warning signals, and such. Admittedly, it is not always feasible to eliminate hazards, as suggested above. However, this should be a listed option, as impracticable as it may seem at first. Keeping this option open requires the issue and possibility to be addressed and, as a result, forces the decision maker to evaluate alternatives that may be available and workable. Situations, circumstances, operations, and business and market conditions change, and what was not feasible last year may be feasible this year.

For less serious hazards and threats, the decision to pursue either the intrinsically safe or extrinsically safe route must be weighed against costs and consequences of alternative actions. Reliability of monitoring techniques must be weighed along with their costs and comparisons of their effectiveness with that of other approaches.

Life-cycle management is principally concerned with life of components during usual operation, material response to various use and environmental factors, and cumulative effects of interacting damage mechanisms that are a consequence of normal service. Life-assessment methodologies are geared to this and are not intended to factor in effects of abnormal and unusual occurrences, sudden overloads, operating errors, natural disasters, and gross deficiencies. These are not to be ignored, but these events, conditions, and their effects should be evaluated separately.

Materials damage: The focus

At some point in the analysis, essential operations or some indispensable activity or piece of equipment will be identified as a principal hazard or site for potential failure. Alternatives may have been evaluated and no practicable option found for eliminating or diminishing the risk. (A reminder: we are not dealing here with original *design* options, but are concentrating upon ongoing operations of existing plants, established facilities, and processes.) It may be a handling procedure, a process reaction, high temperatures and/or pressures, moving machinery, or other hazardous conditions. Hazards analysis, FMEA, or FTA will reveal how these unavoidable hazards may be transformed into failures.

Since our primary interest is in mechanical failures, we are concerned with integrity of materials used in these hazardous operations, whether they are high-temperature or high-pressure vessels, piping, structural members, mechanical drive components, process reactors, or storage tanks. During their use, they have been subjected to an array of damage mechanisms. Which material is affected by which mechanism is not always predictable. And there are inevitable interactions among several possible mechanisms, all occurring simultaneously, over the entire life of the component.

One mechanism may be time and temperature dependent under constant load, like creep and various forms of microstructural embrittlement. Fatigue cracks may propagate from internal voids and cavities resulting from accumulated creep damage. This condition may be aggravated by fluctuating temperature and need for frequent shutdowns and startups.

In other applications, changes in operating conditions or environments may diminish the effectiveness of protective scales. This may be the result, for example, of increased chloride or sulfide contents of combustion fuels. Higher velocities of erosive substances entrained in gas streams may scour scales from selected regions resulting in localized metal wastage. During service, welded regions may become increasingly susceptible to higher corrosion rates due to chromium depletion as a result of unanticipated exposure to sensitizing temperatures.

Over the lifetime of mechanical components, many different and random forces, environments, and circumstances unforeseen and unanticipated by original designers are possible. Any of these, or combinations thereof, can damage materials and set the stage for some other condition or damage mechanism. Anything is possible. Much of this cannot be anticipated by designers. All they can do, really, is to use their best judgment based upon all available information at hand. Then they try to and apply the best techniques and conservative practices in an attempt to compensate for unknowns of the service or use-environment and response of the materials involved. Many vessels, structures, and components still in use were designed without the benefit of recent developments in materials science, fracture mechanics, and damage-tolerant design methodologies, or without an appreciation of the susceptibilities of higher-performance materials to serious damage mechanisms.

Also, the longer the equipment, system, or component has been in service, the more likely that there has been cumulative damage of some kind. But new fabrication is not immune. Residual stresses will be higher. Fresh surfaces, unprotected by films and scales, will be more vulnerable to corrosion and erosion. And the likelihood of design errors, fabrication flaws, and inherent defects will be greater.

When an engineer is faced with the need for evaluating the integrity of a given component and determining its probable useful life or whether it is fit for continued service—and for how long—there is often no way of knowing the original designers' approach, their level of understanding about damage mechanisms, and allowances that they might have made. Were their assumptions overly conservative, marginal, or inadequate? The decision to implement life-cycle management practices usually is made at some time well after equipment has been placed in service. The interval may be 20 years or more. It is important to realize that the methods employed in assessing remaining life are incapable of evaluating the entire system as a whole for cumulative damage. No technique can provide a single clear-cut answer or quantitative indication of how much longer the component or system can be safely operated.

Specific tests provide indications of conditions or their absence at the time of assessment. It is up to the engineer conducting the assessments to make the decisions on system integrity and remaining life. It is wholly a judgment call based upon the integrated assessment of results from all evaluations and tests, the engineer's experience and that of associates and consultants, and input from a great many sources including hazards analyses, failure analyses, FMEA, and FTA.

Evaluation methods simply furnish a "snapshot" of a condition at the time of the evaluation. It cannot tell much, if anything, about the future, although the engineer may be able and should be able to deduce it. The point here is that a single assessment or group of assessments has limited value. It is much like the value of a pilot's single observation of an aircraft's altitude or attitude indicators while flying under instrument conditions. Safe pilotage requires repeated observations and continuous scanning of instruments to enable minor corrections to be made at appropriate early stages. To be effective, implementation of life-cycle management practices similarly must be a continuing activity. Only through repetitive assessments may information be collected, trends observed, and interpretations made in terms relevant to service life.

Evaluation methodologies and scope

Methods for assessing the integrity of existing components depend upon the material, its service and application, and the nature of the operation, as well as historical information obtained from prior experience, analysis of failures, and other factors. Two general approaches are used: operational appraisals and material and condition assessments. Material assessments are often subdivided into nondestructive evaluations and destructive material evaluations. Condition assessments deal directly with the system and monitor conditions that can affect or indicate system integrity. All types are usually needed for accurate analysis.[127–129] Unfortunately, the more useful and pertinent information is usually obtainable only from the more difficult-to-perform evaluations. Each type can be conducted to varying depths of detail. What is done depends to a great extent upon the situation, the findings of prior inquiries, and criticality of the component and its continued safe operation.

All of this requires a game plan. It should address rationale for and purpose of the assessments, continuing availability of funds for this activity, and specific objectives and a general schedule for carrying it out. The plan is subject to revision as experience in the area increases, work progresses, and information is uncovered.

The assessment method chosen for evaluation depends upon the component condition or property of interest, and this reflects back to preceding hazards analysis, FMEA, or FTA. Extent of assessment and technique employed is also determined by the given situation, predominant damage mechanisms, and information needs. These decisions require knowledge of the damage mechanisms encountered in the material and environment, and good understanding of materials behavior and of the related processes and operating details. This can be part of the initial steps.

Operational appraisals

Operational appraisals start with readily available information. Items and characteristics might include the following:

Review of plant history

Component operating records

Available design documents

Applicable codes and standards
Time–temperature profiles
Loading history
Environmental factors
Incident reports
Maintenance and service records
Overhaul or modification information
Manufacturer's recommendations
Production records

From this information, it is possible to determine such things as total exposure time to operating temperatures and operating environments, numbers of loading cycles and shutdowns, and so on, for the system. Knowing the material type and with operating information and available test data, it should be possible to estimate the extent of cumulative damage already sustained and remaining operating life. These are, of course, only rough estimates because the materials data probably were developed in laboratory tests using simply loaded specimens in individual environments and using materials compositions that may be significantly different from those of the actual plant (although of the same type). Fatigue test data similarly might have been obtained under idealized linear test conditions where cycle and stress characteristics, and any number of other variables, differed from those of the actual system.

Nevertheless, it is instructive to perform these calculations. They can provide preliminary indications of whether the component is operating near the end of its projected lifetime or at some earlier stage. Exercises at this stage are not intended to provide definitive answers, merely guidance. But these preliminary inquiries and calculations should be sufficiently detailed to provide a good perspective of the system and component, related operations and characteristics of its probable condition and integrity. They will also reveal information gaps and missing or incomplete details. These form the basis for subsequent study that more directly concerns the hardware itself.

Most often, initial studies are carried out using nondestructive evaluation (NDE) or nondestructive testing (NDT) techniques. It may be found, however, that the most relevant information involves critical components that are inaccessible during operation and are available for examination only during shutdown periods. If this is the case for the component or system of primary interest, these evaluations and others that should be conducted may have to be deferred until that time.

Nondestructive evaluation

A variety of activities and procedures are considered NDE methods. Flaw detection and characterization methods are probably the most common and most critical. Leak detection, dimensional measurements, microstructural evaluations, mechanical and physical property assessments, and stress–strain response are others. A number of nondestructive techniques used in assessing integrity of operational systems and components would be considered monitoring methods; consequently, their descriptions are deferred until the next section, devoted to this.

Methods that detect cracks and crack-initiating flaws are probably of greatest interest in life-extension activity. Detected cracks do not automatically signal a forced shutdown. Virtually every structure or component contains cracks or cracklike defects and flaws somewhere. Their significance hinges upon factors discussed earlier, such as material fracture characteristics, operating temperature, tensile stress level, type of service involved, and the environment. Crack size and shape, crack location, and their growth characteristics are other factors. An objective of crack detection procedures, therefore, is to determine locations, sizes, and shapes, and their implications upon equipment life.

Somewhat different approaches are employed in crack detection (described above) and in crack monitoring (described in the next section, where the interest is in growth trends), although techniques may be the same. Maximum tolerable crack size may be specified for critical high-performance components for such applications as aerospace and military equipment and an increasing number of others. Specification documents may also list crack-detection methods and inspection procedures used, frequency of testing, and other related requirements. In applications where such practices exist and are adequate, they should be followed. Some specification requirements are intended for original equipment manufacture or for use during fabrication or construction. Others are intended for operating equipment.

The field of NDE (or NDT) is vast and there is no intention here to even begin to cover it. Its features are briefly summarized, as are other techniques used in failure avoidance, simply to mention tools that are available. Choice of method for crack detection and evaluation will depend not only upon type of material (e.g., ferrous or nonferrous) but also upon its thickness, crack size that is desired to be detected, geometric configuration of the part, and its microstructural characteristics.

Principal inspection techniques are visual, liquid penetrant, magnetic particle, eddy current, radiography, ultrasonic testing, and acoustic emission; there are others.[130] Standard practices for all of these, and other methods, have been developed by various organizations, including the American Society for Testing and Materials (ASTM), American Society for Non-Destructive Testing (ASNT), the National Institute for Standards and Technology (NIST), and various branches of the military. Important considerations in the use of any NDE technique are its reliability and level of detectability.

Visual inspection. This is the most common inspection method. It is widely used, either alone or in conjunction with other methods. Its only requirements are good eyesight, knowledge of the objectives of the inspection and relevant materials characteristics, and understanding of applicable acceptance standards. It is the primary NDE technique and is used initially to characterize the overall condition of components and systems. It can detect surface flaws and cracks, the presence of corrosion, and other forms of deterioration and contamination, and can evaluate alignment, distortion, and many other characteristics. The visual observer also has other senses, such as hearing and smell, and these can signal abnormalities and problems.

An array of equipment is used to assist in visual inspection and in enabling inspection to be made in inaccessible areas. Light sources, measuring standards and gages, optical profile comparators, magnifiers, and mirrors are basic tools. Extension devices are often required for visual access. These include rigid and flexible borescopes for illuminating and observing internal or obscured machinery and structures.

Imaging sensors are used for remote inspection and include solid-state charge-coupled device (CCD) cameras, television cameras with vidicon and other types of tubes and various kinds of optical sensors. These detect light and are capable of converting it into electrical impulses that are refined and analyzed into forms that are readily interpreted by observers and users. These signals can be recorded and stored on film, magnetic tape, optical disks, and other media for later imaging playback and reference. These devices may be incorporated into manipulators, transporters, and other such devices for gaining access to remote or hazardous areas for visual inspection.

Procedures and materials have been developed to facilitate visual detection of surface flaws, defects, and cracks. They are described below.

Liquid penetrant inspection. These techniques are intended to enhance surface discontinuities and make them easier to detect.[131] Their application is limited to surface flaws in solid, nonporous materials having relatively smooth surfaces. Liquid penetrants are formulated to provide good wetability and low viscosity to readily seep into cracks and tight crevices and other surface discontinuities where, after suitable processing, they render flaws more visible. These techniques are usable on any material, including ferrous and nonferrous metals and alloys, castings, and nonmetallics such as plastics and ceramics. The materials being evaluated must be at or near room temperature to avoid boiling and decomposing the penetrants.

Two principal types are in use today: fluorescent and visible dye. Fluorescent penetrants glow brightly under ultraviolet illumination and clearly indicate cracks and other surface flaws. These are available in several levels of sensitivity for different inspection requirements.

Dye penetrant is usually red in color. In use, the penetrant is applied to the surface and allowed to seep into flaws. After an appropriate interval, or dwell time, the surface is wiped clean and dried to remove traces of surface dye, then a "developer" is applied to the surface. It leaves a white powdery residue when dry. Red dye that has penetrated surface flaws bleeds into the white residue, making their presence, size, and location visible.

Both types of penetrants are capable of field use with portable equipment and with penetrant materials supplied in kits of pressurized spray cans. When used in accordance with recommended procedures, these methods are reliable, rapid, easy to use, and inexpensive. Their principal drawback is that they are usable only on open and clean surface defects in nonporous materials.

Magnetic particle inspection. Magnetic particle inspection is used for detecting flaws at or near the surface of ferromagnetic materials.[132] The method is comprised of magnetizing the part to be inspected, applying finely divided magnetic particles to the surface, and observing and interpreting the result.

In principle, lines of force in magnetic materials are interrupted or distorted by material discontinuities. These produce flux leakage such that magnetic particles subjected to these conditions tend to accumulate in patterns characteristic of the discontinuities. These serve to reveal their locations, general sizes, and shapes. The magnetic particles may be used dry or in wet suspensions in oil or water, with appropriate additives. They also may be treated with special pigments to fluoresce under suitable illumination.

Although the method is not limited to detection of surface flaws, identification of subsurface discontinuities depends upon their size and depth, as detectability diminishes with decreasing flaw size and increasing depth below the surface. Even cracks tightly filled with corrosion product or other debris may be revealed by this method. Most parts of moderate to heavy mass require magnetization using electromagnetic power supplies and magnetizing heads of various configurations suited to the geometry of the part. In the detection of tight linear subsurface flaws, this method is reportedly superior to radiographic techniques.

Eddy-current inspection. Eddy-current, or electromagnetic, examination techniques are based upon principles of electromagnetic induction. Impedance of a spirally wound conductor changes when it is brought near metallic objects or electrically conductive materials. These changes occur through interactions between the electromagnetic field produced by the eddy currents induced in the part and that of the spirally wound conductor.

In testing by this method, eddy currents are induced in the part being examined. Changes in those currents are produced by changes in the part in electrical conductivity (and resistivity), magnetic permeability, and in metallurgical conditions such as heat-treated condition, hardness, and grain size, as well as defects such as cracks, voids, and other discontinuities.

Induction coils may be a solenoid (cylindrical helix) type or flat helix (pancake). Since no direct contact between the part and induction coil is necessary, the method is well adapted to mechanized production applications for inspecting welded pipe, tubing, bars, billets, cable, and other magnetic or nonmagnetic product forms.[133,134] It is a versatile NDE method that can also be used to measure dimensional variations.

Since it is sensitive to conductivity variations, it can be used to detect compositional differences and has been used in measuring coating or plating thickness, in sorting metals, and in detecting the presence of foreign materials as, for example, in ensuring the absence of extraneous compositions of core wire during welding electrode production.

This is the principle of operation of coin discriminators (i.e., slug rejectors) widely used in vending machines and other coin-operated devices. It is the reason why U.S. coins (dime, quarter, and half-dollar denominations) are laminates of a copper–nickel alloy (outer layers) and pure copper cores. This material mimics the resistivity–density properties and related response in eddy-current coin discriminators of the 90-10 silver–copper coinage alloy that had been used by the U.S. Mint in these denominations of coins prior to 1965.

Radiographic inspection. In this inspection technique, the part to be inspected is placed between the source of radiation (x ray or γ ray) and photographic film. As a result of differential absorption of the penetrating radiation, the film image density varies inversely with the density of the part, with voids, holes, cracks, and gaps being regions of higher radiation intensity and therefore observable.[135]

Optimal sensitivity is obtained when planar defects, such as cracks and delaminations, lie with their planes oriented parallel to the direction of transmitted radiation. Deviations from this "preferred" orientation diminish the detectability of these kinds of discontinuities. Tight cracks with their crack planes oriented normal to the radiation direction are often undetected by this method.

The nature of radiography requires access to both sides of the part being inspected for film placement. This can limit application of this method for some components and part configurations. For some forms, such as tubing and pipe, for which film placement within the internal cavity is impossible except near the ends, radiographic inspection is still a viable method through use of double-wall techniques. Circumferential welded joints in tubing and piping may also be inspected using these techniques by radiographing at an angle to the tubing axis, revealing a nearly complete weld periphery of elliptical shape in the processed film.

Sensitivity and resolution for a given combination of radiographic conditions and material are determined through use of penetrameters. These small waferlike materials have absorption characteristics similar to that of the part being inspected and are placed within the radiographic inspection field, usually in the least favorable region. A penetrameter thickness of 2% of part thickness is typical. Many penetrameter designs are used. Their purpose is to evaluate image quality. Usually this is expressed in terms of the size of the smallest feature that is visible in the processed image, such as a hole diameter or step-thickness variation (e.g., a "2-*T*" sensitivity indicates that a hole in the penetrameter wafer having a diameter of twice its thickness is visible in the processed film).

Interpretation and acceptance standards for evaluating the radiographic quality of the part have been prepared by many organizations, industry associations, and government agencies. Designation of such standards usually falls within the scope of product specifications and therefore is a matter of contractual agreement between vendor and purchaser. They describe maximum permissible sizes for flaws of various types, orientations, and groupings, as a function of part thickness. Some formats use charts, schematic diagrams, or reference radiographs to assist in classifying flaws and their degree of severity for various categories of service criticality. The extent of radiographic inspection is also described. It will indicate whether a full or partial inspection is required and sampling requirements for production situations. Procedures for reinspection after defects have been found and repaired are also spelled out.

Ultrasonic testing. Ultrasonic testing or inspection involves directing a series of high-frequency sound impulses into a test part. When the pulses impinge upon obstructions or the other side of the part, which are parallel or nearly so to the pulse initiation surface, they are reflected back. These reflected signals are detected by the transducer and can be amplified and displayed on the screen of a cathode-ray tube (CRT). The same electronic equipment transmits the original sound pulses and detects their echoes.

In the pulse-echo straight-beam mode, the time interval between signal bursts or pulses permits the reflected sound pulses to be returned to the transducer and processed. Since sound propagates in solid materials at predictable rates, this characteristic can be used as a basis for determining distances to echo-producing flaws and discontinuities. Through comparisons of amplitudes of displayed discontinuity signals to those of reference standards, both location and size of the flaw may be estimated.[136]

Calibrated test blocks or reference standards containing artificially produced flaws (e.g., flat-bottom holes or precracked implants) are used in presetting test equipment and in characterizing echo sites and in interpreting test indications. Test block materials are selected to have sound propagation properties similar to those of the part being tested.

This method is most sensitive to delaminations or cracks whose planar surfaces lie normal to the direction of the propagating sound. Note, in this respect, that the method is most sensitive under conditions where radiography is least sensitive. The technique is capable of detecting both surface and subsurface discontinuities—even deep within thick members—and flaw size and location may be determined quantitatively.

Ultrasonic inspection provides higher sensitivity and greater accuracy than other NDE methods in identifying and characterizing internal discontinuities. Testing equipment can be portable and self-contained and the process is not hazardous to use, in contrast to radiography. These techniques are widely used in many industries for detecting the presence of damaging flaws in critical components.

Acoustic emission. This inspection method utilizes stress waves produced in stressed materials by microstructural events, including cracks, and other material movements such as deformation. In use, acoustic piezoelectric crystal sensors detect stress waves emitted from the material elastic stress field which are amplified and processed in the detector and measurement circuitry and displayed in appropriate formats or recorded and stored for later retrieval and playback.

Acoustic emission (AE) inspections are typically conducted during controlled loading or "stimulating" of the structure or component being evaluated. The loading may be some controlled operating configuration typical of service experience, a proof test, or a fatigue or creep test.

A notable feature of AE is that, with suitably located sensors, the entire component or structure is evaluated all at once during a single loading or test session. It is a nondestructive and nonintrusive method and can be carried out with the equipment being tested in place with no need for immediate inspection access other than to affix the sensors. Its principal advantage is its ability to signal potential problems that might exist throughout the entire unit. With these identified and broadly characterized, other methods can be applied to delineate more closely the defect or detected acoustic event.

AE equipment is sensitive to signals of all sources within its frequency range; therefore, precautions must be taken to minimize obscuring effects. Also, well-trained personnel are required in setting up and conducting these inspections and in interpreting results. Standard inspection procedures have been developed by various organizations for a wide range of materials, including nonmetallics such as fiber-reinforced plastics.[137,138]

Significance and application of NDE results

As noted before, characterization of flaws in terms of component properties can be even more significant than their actual detection. From fracture mechanics principles, a key issue in the use of NDE as a failure avoidance tool is critical crack size under operating conditions for the material or component. Related issues relate to growth characteristics for existing cracks and steps for preventing catastrophic failure.

This boils down to detectability, in given components, of critically sized flaws, their probability of detection, and level of confidence. This will vary with flaw shape, location, and material.[139] Of equal and often greater importance to life-extension issues is knowledge of crack-growth characteristics and detectability of cracks that may grow to critical size during inspection intervals.

These are most important factors, as they determine the success of avoiding catastrophic failures.[140] However, the lack of precision in determining crack sizes and crack-growth rates in actual components and structures using state-of-the-art techniques cannot be overstated. Advancements have surely been made, but much more needs to be done. Use of a number of complementary approaches significantly increases the probability of identifying problems at an early stage and thereby more effectively avoids failures.

For operating systems and components representing some level of cumulative damage, continuous monitoring offers the most relevant information for failure avoidance. Principal techniques are described in the following section.

Real-World Monitoring

Importance of response profiles

The preceding section described operational appraisal procedures for estimating integrity of a system or component. This is done through observations and information that would offer a basis for judging the degree of cumulative material damage sustained during service life to date. Because components and structures differ widely in their materials, service demands, and environments, it has not been possible to deal in specifics. However, evaluation procedures have been presented and discussed which apply to most situations. As necessary as these procedures are in initial appraisals, they are of an indirect nature and therefore provide only limited insight on component life. There are many potentially serious damage mechanisms and combinations, as already noted, and initial operating appraisals may not reveal their presence, or the extent to which they may have already progressed in the component, structure, or system of interest.

The core of a life-extension program, however, comprises ongoing material assessments and condition monitoring procedures. Some are direct; others indirect. Their primary purpose is to provide a continuous record of response to real-time conditions of some aspect of components or systems that are most at risk. The indicated response should be in terms that are significant to the integrity and continued satisfactory operation and performance of the component or system being monitored.

Avoidance of catastrophic failures is always a paramount concern, and the role of cracks and crack-involved damage mechanisms has been emphasized. However, many kinds of subcatastrophic failures can occur that are devastating and can have seriously adverse effects upon a company's viability. We should not become so preoccupied over avoiding catastrophes that we neglect apparently trivial causal elements that can lead to other types of failures that also must be avoided. *All* hazards identified in the analytical studies discussed earlier require attention.

During our discussions of failure analysis we noted that its value was due largely to factual information it provided on response to real-world conditions. In some respects, this experience supersedes results of laboratory tests that attempt to determine response under simulated service conditions. But, as we said, as valuable as failure analyses are, we cannot wait for failures to occur to gather information on how to extend the life of operating equipment.

This is where monitoring comes in. If it is done properly using appropriate techniques and conducted in a consistent manner, and if the results are carefully studied and analyzed, monitoring can be a most effective tool in avoiding failures—of all kinds. The monitoring need not be conducted full time, around the clock, but may be periodic and done at predetermined intervals. Some monitoring procedures cannot be carried out while the system is operating due to accessibility restrictions, high temperatures, hostile environments, or other reasons, but can be conducted only during scheduled outages or when the equipment is turned off. However, the principal value of monitoring lies in providing a continuous response profile over an extended period of time.

Monitoring can have a variety of formats. It can be nondestructive and measure some property or response of material or operating system condition. It may be confined to measurement of some parameter or detection of some condition, or it may be used to observe changes or rate of change in a component. Its methods may vary in their level of sophistication from simple measurements of temperature to more complex techniques of signature recognition, and everything in between. Its purpose usually is to provide a warning, and awareness capabilities and requirements are important considerations in the choice of methods used. Other purposes would include data gathering or condition documentation to furnish input to other programs. Decisions on what to monitor and its requirements should be based upon hazards analysis and other analytical procedures such as FMEA and FTA.

Monitoring can also involve so-called destructive tests where actual samples of the component are removed and evaluated. These are direct types of evaluations. Monitoring can also be directed at indirect effects or by-products of the operation such as measurements of pH of effluents or other levels of toxic substances. Or it can involve measuring and observing the response of test specimens exposed to the same operating environments of actual systems.

Condition assessments

Visual monitoring. This may seem to be an obvious method, but it is the dominant one and its significance cannot be overlooked. Aside from the usual procedures of visual inspection discussed previously, other techniques can make visual monitoring a significant tool in condition assessments.

Television techniques, with their capability for converting light impulses into electronic signals, are ideally suited to visual monitoring and recording. Remote controls, an array of lenses, image intensifiers, improved focus controls, high definition and resolution, and color capability have recently expanded their scope and utility. Monitoring of inaccessible locations such as underground, underwater, and hazardous sites, or within vessels, pipelines, and confined areas is now possible using these devices. They can monitor environment, operating conditions, and integrity of systems and components which, until only a few years ago, were literally out of reach of any monitoring technique.

Stereophotography. This is an increasingly useful technique for remote evaluation of the condition of equipment and materials. Because of its capability for three-dimensional depiction of photographed objects, it is especially adaptable to monitoring surfaces affected by corrosion, erosion, and wear. Since the method can be used remotely and provides a permanent record, this technique should find expanding applications in failure avoidance activities.

Stereophotography is also helpful in examining and interpreting fractures and other surfaces subject to various forms of material degradation. Methods have been developed for producing stereo pairs during scanning electron microscopy (SEM), and these provide the examiner with a three-dimensional, high-resolution view of surfaces of interest.[141]

Optical holography. Along similar lines, laser science has led to development of optical holography and its capability of reconstructing three-dimensional replicas of actual components. It has particular usefulness in analyzing strain and vibration behavior of materials and for depth-contour mapping and other nondestructive inspection and monitoring tasks.[142]

Photography. Photography, in both still and motion picture formats, has been a principal method of visual monitoring and continues to be. Image quality and resolution are generally superior to those of other methods, and a large selection of films and emulsions for special applications is available. Photographic formats are easily stored and retrieved and may be converted to electronic media by recently developed techniques.

Motion analysis of moving objects is readily accomplished using high-speed photographic techniques, and can be an indispensable aid in identifying causes of failures in moving machinery and production situations.[143]

Leak monitoring. Leaks of gases, liquids, and vapors have been the source of numerous catastrophic failures and account for significant damage and economic loss. Monitoring of equipment to detect leaks can prove most useful in avoiding these failures. Besides this, leaks can be indications of impending loss of system or component integrity. And leaking substances can cause unanticipated deterioration of materials that they contact and contaminate.

Gas and odor detectors have been developed for many gases using various principles ranging from simple to sophisticated. Principles of operation include chemical reaction color changes, catalytic heating, flame ionization, thermal conductivity, gas chromatography, mass spectrometry, and others. Leak-rate measurements can be made using tracer gas, which reacts in a predictable way with leak detection equipment. Gas concentrations achieved over a given period of time are measured and used in determining leak rates.[144,145]

Continuous leak monitors for toxic gases, liquids, or explosive mixtures are equipped with alarms set at low detection levels. These are routinely used for processes involving ammonia, carbon monoxide, hydrogen, methane, and other hydrocarbons, including smoke.

Communication cable integrity is commonly monitored using cable pressurization techniques where a small positive pressure of a dry nonreactive gas is maintained within the sheathed cable. Leaks and their magnitude are detected from pressure transducer indications that are transmitted periodically to central offices and monitored frequently.

Acoustic monitoring devices that are sensitive to sound frequencies characteristic of gas and liquid leaks have been developed and are available. These may be connected to alarms or other readout devices to signal leaks of systems, vessels, boiler tubes, pipelines, or other components. Similar devices are used in intrusion detectors and are tuned to specific sound frequencies characteristic of breaking glass or other intrusion-generated signals.

Vibration monitoring. Vibration monitors are used in detecting and characterizing changes in system vibration modes which may be symptomatic of impaired structural integrity, loss of operating tolerance (through wear, for example), or failure of components. Vibrational changes can also lead to initiation of fatigue cracks or accelerated growth of existing cracks and defects. Accordingly, vibration monitoring and analysis are issues of major concern to users of mechanical equipment subject to fatigue failure.

Selection of monitoring subjects and locations should be based upon susceptibilities identified in hazards analysis, results of failure analysis, and other information. An understanding of how abnormal operation or system problems may be translated into changes in vibrational frequency or waveform is also essential.

Initial assessments of suspected problems can often be made visually using stroboscopes to view moving machinery over its range of normal operating speeds. Abnormal displacements can sometimes be spotted using this technique and can indicate probable causes for suspected difficulty.

Vibration monitoring requires baseline information on "normal" operating modes to serve as reference "signatures" for the specific system. These are established during normal operation of the equipment or system, throughout its various operating modes and cycles. Deviations from this baseline behavior observed during monitoring of subsequent equipment operation are analyzed and interpreted for system reliability implications. Such deviations could signal the onset of fatigue, overload, or some other hazardous condition. Interpretation of these indications usually requires additional information on other operational or environmental variables.

In critical components, the source of vibrational anomalies should be promptly identified and corrected before resuming operation, as they may be indicative of a structural problem, a loose connection, failed weld, or other potentially serious condition. Continuous monitoring can also reveal gradual changes in signature and trends taking place that can indicate progressive deterioration. Timely corrective action can forestall major repairs and extended downtime.[146,147]

Substance condition monitoring. Operational monitors may be supplemented with condition monitors for fluids and other substances used in or associated with the operation. These indirect methods often serve as early warning indicators of problems not detectable otherwise, or until failure incidents occur.

For example, spectrographic analysis of lubricating oils within closed circulating systems, as in aircraft engines, can be helpful in evaluating lubricant quality and detecting abnormal wear, overheating, presence of contaminants, exposures to adverse environmental conditions (such as dusts, chlorides or sulfides, and moisture) and other potentially detrimental effects. These analyses should be conducted on samples taken at periodic intervals to provide a basis for observing trends, as these are often more significant than individual test results.

These analyses usually give particular attention to metallic elements that are characteristic of specific engine components. Since aircraft oils are compounded to be free of metallic additives, traces of metallics can be suspected to have originated in engine wear or contaminants. Often, the location of abnormal wear can be identified and assist in avoiding costly repairs due to complete breakdown or even more tragic consequences. Analytical results are furnished with values for the various elements typical of normally operating engines of the specific type and model.

Similar analytical monitoring surveys of coolant, exhaust gases, other lubricants, and effluents associated with the operation or process can provide helpful clues in forestalling failure. These evaluations can detect depleted additives or corrosion inhibitors, chemical imbalances, excessive levels of pollutants or contaminants, corrosion and material breakdown, pH changes, wear, and other system characteristics and abnormalities.

Integrated monitoring systems. Traditional monitoring of operating systems measured temperatures, pressure, flow rates, rotational speeds, and similar conditions. These were visually monitored and manually recorded in operating logs, and operator judgments were made and actions taken accordingly. Abnormalities and malfunctions were detected largely by human senses, and system integrity depended heavily upon operator experience and knowledge of the equipment or process.

With the advent of improved diagnostic instrumentation, computerized data links and information storage and retrieval capabilities, and interactive graphic displays, integrated monitoring systems are possible. These can offer system operators a more accurate picture of on-line real-time operating conditions, and communicate changes that signal abnormal behavior and degraded performance of some component or subsystem.

A prototype or test-bed system recently installed in an electric utility generating station through a joint effort by Philadelphia Electric Co. and the Electric Power Research Institute (EPRI) contains acoustic leak detectors to give early warning of boiler tube and other leaks.[148] Acoustic gas temperature monitors are installed in exit gas lines, and steam temperature monitors at each heating section can reveal decreases in heat absorption due to slag buildups. Ultrasonic transducers monitor turbine bearing wear, vibration analyzers check for rotor dynamics, and stress analyzers measure stress in the boiler wall and in the turbine casing and rotor. Acoustic blade monitors detect excessive blade flutter, and numerous thermocouples monitor component temperatures throughout the system.

Since 1958 in the United States, similar systems have been used in commercial aircraft flight monitors ("black boxes"). These are dedicated to monitoring information and its storage only and have no feedback function during data collection mode. They represent a complex array of sensors and instrumentation links and are the collecting point for principal flight and engine data, plus a log of navigation events and decisions. These monitors frequently provide aircraft accident investigators with their only clues to causes of aircraft crashes. Their retrieval and evaluation can be key elements in the investigation of incidents involving commercial aircraft or failure of its systems.

Such monitoring configurations might be considered elaborate for operators of equipment or facilities of lesser magnitude than electric power-generating plants or complex commercial jet aircraft. However, they indicate the range of possibilities for not only forestalling major failures, but also in minimizing downtime and in more efficient scheduling of maintenance. Monitoring systems can be implemented for any critical operation and may be custom-designed to satisfy specific needs and requirements.

Material assessments

In conventional NDE techniques described in the previous section, actual components were evaluated for flaws and other conditions that could lead to cracking and eventual failure, but the materials themselves were generally unaffected, that is, the evaluation methods were, to some degree, remote. Most nondestructive material monitoring methods have more intimate involvement with the material.

Traditional NDE methods are intended primarily for detection of existing cracks and flaws, but are unable to detect incipient damage. Effective failure avoidance requires identification of conditions that are precursors to crack initiation. Many techniques have evolved over the years and are well developed. Others have been recently devised and are undergoing field tests. Some of the methods and concepts are new and largely unproven; however, they are included here as examples of approaches that are being considered to provide better insights into the status and integrity of operating components.

Strain monitoring. In the field of materials science, strain is defined as the relative change in size or shape of an object under mechanical load. Under ordinary circumstances, the term refers to linear, or longitudinal, strain occurring in the direction of applied stress. Since linear strain is expressed as a change in length over a given distance, the term is dimensionless (i.e., in/in or mm/mm).

Results of conventional (uniaxial) tensile tests are depicted graphically as stress–strain curves, where stress (force per unit area) is plotted against strain (deformation per unit length). For some materials, including metals, strain is proportional to applied stress (elastic behavior) up to some point termed the elastic limit or proportional limit.

Stress, force, torque, and pressure sustained by a part are frequently important quantities to measure but cannot be measured directly. But, because of known stress–strain relationships for specific materials and the ability to measure dimensional change in parts under load, stresses can be readily calculated. Measurements of strains and determinations of their magnitudes, directions, and locations are tasks of stress analysis.

Because of high deformation resistance of metals, strains occurring in components under usual loads can be very small, requiring precision measurement capability. For example, precision dial indicators mechanically affixed to tensile test specimens were used for many years in measuring tensile strain. Such delicate mechanisms are impractical for most strain measurement applications involving operating equipment and have been generally replaced by bonded electrical resistance or semiconductor strain gages or transducers.

Operation of these strain gages is based upon the property of conductive materials in which electrical resistance varies with changes in length under stress. To obtain accurate strain measurements and to minimize spurious gage effects, resistance strain gages are made extremely thin and small. The strain-sensitive electrical resistor element, having a gridlike pattern, is bonded to backing material that is easily attached to the component to be measured. Strains occurring at the surface of the component are transmitted to the gage element, resulting in changes in electrical resistance. These resistance changes are converted by the detection circuitry into direct readouts of strain, stress, pressure, or other desired quantity.

Strain gages measure strains in the axial direction of the grid and are designed to be insensitive to transverse strain. The use of a single strain gage, therefore, provides information on uniaxial stress states only. Consequently, their application is confined to situations when direction of the principal stress is known (as in a laboratory tensile test). For measurements and characterization of biaxial stress states, other multigage rosette configurations are used. These permit a sufficient number of strain measurements to be made to evaluate their magnitudes and directions. Although shear strains are not directly measurable with resistance strain gages, gages with specially designed shear-pattern configurations provide measurements that can be used in calculations to obtain shear strain data.

Strain measurements are fundamental to evaluating stress states and loading configurations of components, structures, or parts. Many applications of interest involve operation in hostile environments and at high temperatures. Special electrical resistance gage materials have been developed that are capable of withstanding temperatures up to about 400°C (750°F) for brief periods, and capacitance-type gages are reported to have performed satisfactorily at 600°C (1112°F) for up to 20,000 h.[149] Effects of temperature, humidity, prolonged exposure to adverse environments, and other degrading influences should be taken into account in selecting gages, adhesives, lead connectors, and such for field applications.

Since electrical resistance strain gages are unsuitable for use at elevated temperatures where it is desirable to measure strains to detect creep damage, other methods are being explored to supplement currently used dimensional measurement techniques. Dimensional measurements reveal overall effects whereas the new methods are intended to measure more localized creep strains. Of particular interest would be locations of stress concentration or where microstructures are less creep resistant.

One approach being studied involves scribed grid patterns on the surface of the component.[150] They are protected during operation by suitable surface coatings. Replicas made of these grid patterns serve as reference information on the biaxial strain state of the component at the grid location. Comparative strain evaluations of replicas made by measuring the grid configuration at spaced time intervals (e.g., between scheduled shutdowns) can provide strain-rate information.

Other methods used for strain measurement of components operating at high temperatures function in much the same way, although they may not offer indications as localized or as accurate as the above method. These usually involve taking precision measurements between markers permanently affixed to the structure or component. The markers may be welded-on projections or indentations. They are often placed within regions that have been weld overlayed with nickel-based or cobalt-based alloys to prevent oxidation and scale buildup from obscuring their locations.

Measurement between markers is made to typical micrometer accuracy (± 0.001 in or ± 0.025 mm). They are usually capable of identifying strain rates of 10^{-8} per hour for maximum total strains of 5% to 10%.[151] Choice of locations should be based upon criticality considerations and measurements should cover regions of greatest concern for creep damage. These usually encompass welds because of their typically inferior stress-rupture properties and in which grain boundary cavitation can occur during shorter high-temperature exposures than in unwelded wrought material. These types of strain measurements across relatively large distances are, of course, averages and may not faithfully represent localized situations, which may be controlling. These observations across relatively large distances should therefore be supplemented with information from other examination methods.

Sometimes it is desirable to obtain information on the stress state of an entire part or component, possibly for guidance in properly locating resistance strain gages. Strain gages, by their localized nature, are unsuitable for this. Several techniques are available for such evaluations. One is the *brittle coating* method.

It employs a highly strain-sensitive lacquer that is used to coat the part in its unstressed state. After the part is dried and stressed, the lacquer coating develops cracks running normal to the direction of maximum strain, somewhat like a candy apple. The technique not only offers an overall perspective of strain distribution but also its direction and magnitude. It is a useful technique for assessing moving or inaccessible parts that cannot be instrumented and provides general qualitative information on locations of critical stress.[152]

Photoelastic techniques provide similar, although more accurate, information on overall strain behavior within a part. Some variations of this method employ transparent plastic replicas of the entire part, but strain behavior of metal components is also evaluated through applications of coatings of photoelastically sensitive transparent materials. Under stress, these materials become birefringent, resulting in a display of color fringes when viewed under a reflection polariscope. Surface strains are apparent through observing fringe patterns and spacings.[153]

Other strain measurement techniques use combinations of methods. Stress relaxation techniques usually involve material removal and therefore are not NDE methods. They provide reliable quantitative data, however, and are therefore commonly used.[154]

One practice of this kind in evaluating residual stress in plates or similar two-dimensional members is to affix electrical resistance strain gages to a location, then remove a wafer of material (containing the strain gage). This is done by machining a circumferential groove around the location to release the strain-gaged "island." After its removal, changes in strain in the removed section are measured with the gages still in place. From this, the residual stress pattern in the original plate can be calculated. Although a major drawback to this method is that it is destructive, necessitating weld repairs of the site, the quantitative information it offers is often regarded as outweighing its disadvantages.

X-ray diffraction is applicable to crystalline materials (this includes metals) where elastic strains can be measured using the lattice parameter and is a nondestructive method. Other techniques, including ultrasonic methods, are under development and are based upon measurements of the stress-sensitive properties of metals.

Surface metallography. In this technique, standardized practices used for decades in preparing metal samples for laboratory microstructural examinations are extended to field use and in situ examinations of operating equipment. Although the several material preparation stages involve removal of superficial layers of material, the amounts of material removed are so trivial that it is considered a nondestructive evaluation technique.

These methods have been used in life-extension activities mostly in the fossil-fueled power-generating industry. Their objectives have been to monitor microstructural features and effects, with emphasis upon microcracking and creep cavitation, conditions relevant to the service life of these components. However, these techniques are well suited to similar evaluations of any operating equipment in any industry and are being increasingly applied elsewhere.

Two general methods are used. In one, the prepared surface is examined directly using a metallographic microscope. Such examinations are usually limited to horizontal surfaces due to microscope positioning and stability problems in other configurations. Most metallographs are equipped with photographic capability with builtin illumination. This permits photomicrographs to be made at various magnifications (up to 1000 $\times$) of regions of interest. The regions are previously prepared by grinding, polishing, and etching to reveal desired microstructural features.

The second examination method, and the one most widely used for these purposes, involves an intermediate step of preparing a cellulose acetate replica of the metallographically prepared surface. Any number of regions may be prepared and replicas made of them. These may be taken to the laboratory for a more comprehensive examination than is possible in the field and under more favorable conditions. Besides, replicas are tangible reference material that may be filed for subsequent comparisons and study.

Replication techniques can be applied to any reasonably accessible surface. They are not subject to the difficulties inherent in attempts to conduct high-resolution examinations in the field due to dust, vibration, and other environmental contaminants. Field examinations of prepared regions are necessary to verify that they cover subjects of major interest. Also, completed replicas should be examined at the location to ensure that they are of the proper quality. When several replicas are made and different locations are involved, a reliable identification system is essential. A photographic record showing replica locations designated on the actual equipment is also needed for later orientation and documentation purposes.

Since replicas are negative images, interpretations are often more easily derived if they are viewed in the SEM using reversed-contrast imaging techniques. Alternatively, reversing film can be used to obtain positive images from the replicas. Thin sputter-coated layers of a conductive material, such as gold or gold–palladium alloy, are necessary for examination in the SEM and will improve optical microscopy examinations as well.

Selection of appropriate locations on the component or equipment for metallographic examination is an important decision in the use of these techniques in life-extension activity. Regions identified as critical for both metallurgical and service-related reasons may be highly localized. Since these regions are more susceptible to damage than other regions, they require priority attention. Knowledge of materials response under the equipment's operating conditions and an understanding of localized effects of temperature and peak stresses are essential in identifying regions for examination. This knowledge may be obtained through observations of failure modes of laboratory-tested material or through failure analysis examinations of similar materials from similar equipment and service conditions.

Observations of material damage in the replicas may be correlated with appearance of test specimens at various stages of life or accumulated damage or with other materials from actual components to determine damage level and lifetime stage represented. They also may be used in pinpointing critical regions for further study using other inspection techniques.

Somewhere in every operating system there is a region of maximum accumulated damage. If the area can be identified, it can serve as a critical "front-line" sensor upon which to base subsequent life-extension assessments. Also, if it represents the "weak link," the region can be modified to decrease peak stresses, high temperature, or other service or structural changes made. Prior operating history, knowledge of materials response under operating conditions, experience in examining samples from this and similar equipment, and input from other sources can assist in these evaluations.

A limitation to metallographic techniques as described so far is that they are confined to outer surfaces. Interpretation of their results, therefore, should recognize this, as it may not be safe to assume that surface microstructures are typical of those throughout the thickness of the component or at the inner wall (of tubes, for example). Temperature and stress gradients may exist across the thickness, environmental effects may have degraded an outer more accessible surface, than an inner one, and vice versa, and compositional variations may exist through the thickness.

All of these possibilities can affect the microstructure and its response. If there is a need to examine these variables more closely, samples are often cut from actual components for more intensive laboratory examination. These are usually small boatlike segments, slices, or trepanned cylinders, and can provide more complete information than surface replicas. Since they are destructive to some extent and require repairs, there is an obvious limit to the number of such specimens that can be taken. Critical regions, where section thicknesses change, fittings are welded, and heat-affected zones of welds, usually are the most difficult regions to sample or prepare for replication.

It should be apparent that these techniques, like so many others, do not of themselves provide clear-cut answers in terms of remaining operating life. They do, along with indications from other studies, furnish the equipment operator or decision maker with ingredients for making these assessments. Obviously, the more facts that are available and the more reliable and relevant the data and indications from various evaluations and other sources, the more enlightened the decisions will be.[155–157]

Monitoring environmentally induced cracking. The severity of materials damage as a result of environmentally induced cracking (i.e., stress-corrosion, hydrogen embrittlement, and corrosion-fatigue) and its consequences call for detection methods that can alert equipment operators to this condition well in advance of complete fracture. Unfortunately, such a capability does not exist. In fact, mechanisms responsible for crack initiation and propagation are not that well understood, decreasing the likelihood of imminent developments along these lines.

All that can be done is to monitor components and equipment for cracks using NDE methods already described. Also, since it is known from observations and experimental results that various operational and environmental conditions promote these kinds of damage, these conditions can be monitored in attempts to avoid potentially damaging situations. This is not altogether practical, however, as stress levels and environments that produce cracking are very often normal design conditions. Regions of known high peak tensile stress and locations of stress concentration can be monitored, as well as environments for ions known to trigger cracking (e.g., chlorides for austenitic stainless steels, ammonia for copper alloys, etc.). These measures can be helpful but are indirect and do not indicate remaining life or how near the component might be to the onset of these forms of cracking.

Microstructural condition often is a principal factor in development of this kind of damage, and this may serve as an indicator of increasing susceptibility if not a sensor of these damage mechanisms. From failure analysis it is known that certain microstructural changes promote stress-corrosion cracking and hydrogen embrittlement. For example, grain boundary segregation of trace impurities, localized depletion, and inhomogeneities (banding, e.g.) cause stress-corrosion in low-alloy steels. Sensitization in stainless steels is another example. Hydrogen embrittlement of iron and nickel alloys has been traced to sulfide segregation. Temper embrittlement in low-alloy steels is known to increase susceptibility to hydrogen-induced cracking and stress-corrosion cracking due to phosphorus segregation.[158] These are the more well known examples, and there are others that are alloy-specific.

In recent years, there have been increasing suspicions among metallurgists and materials scientists working in this field that it is not the high strength level of hydrogen-cracking-susceptible materials per se that leads to these problems, but microstructural strengthening mechanisms themselves that are probably controlling.[159] If this is so, known microstructural effects offer opportunities for development of monitoring procedures capable of identifying the presence of these conditions, thereby indicating susceptibility to environmentally induced cracking in these materials.

Some progress has been made in this direction. For example, hydrogen-induced cracking in electric-resistance welded low-alloy steel oil and gas pipeline systems exposed to sour (hydrogen sulfide) gas was found to be associated with inclusion stringers and pearlite banding (a microstructural feature related to localized segregation of alloy constituents).[160–162] Metallographic techniques discussed above can be applied to materials known to form susceptible microstructures and examinations conducted. While these may not reveal cracks, they can reveal the presence of susceptible microstructural features, permitting corrective steps to be taken and/or further study to characterize the material more completely.

In addition to such techniques, work has been under way to attempt to accommodate environmental effects in design codes for component lifetime prediction, which are generally formulated strictly upon mechanics considerations.[163] A modified approach has been suggested that incorporates integrated mechanistic modeling predictions with real-time in situ crack-growth monitors and environmental sensors.[164] The in situ sensors monitor cracking in actual or reference components and provide continuous feedback on effects of plant operational conditions. This input, coupled with crack-growth-rate algorithms derived from mechanistic models of environmental crack advance, would reflect both fabrication (microstructure, material, and stress profile) and operational (environment, temperature, and vibration) factors.

Primary emphasis for this approach has been upon nuclear system components in light-water reactors, but the range of applications envisioned includes fossil-fueled plants, steam and gas turbines, nuclear waste containment systems, and deaerators. Preliminary results are reported to provide useful guidance in improving stress-corrosion cracking resistance through better water chemistry control and have demonstrated the importance of better definition and control of other factors such as residual stress and grain boundary sensitization.

Preliminary studies have also been conducted on acoustic emission techniques for monitoring stress-corrosion cracking. In laboratory tests, stress-corrosion cracks have been shown to emit acoustic signals, and correlations have been made between the crack area and number of acoustic emission events. According to its researchers, on-line (real-time) monitoring using this principle will require improvements in the ability to discriminate between background noise and crack-growth signals.[165,166]

Corrosion monitoring. In addition to the need for monitoring susceptibility of components to environmentally assisted cracking, there are needs for monitoring conditions that can lead to other types of environmentally related materials degradation or corrosion. A number of approaches of varying sophistication have been devised.

Probably the most widely used approach has been to suspend test coupons within the operating environment. These may be used in virtually any operation and can include machined, welded, prestressed, heat-treated, and other types of configurations. They are prepared of the same materials as operating components, and, since they are subjected to the same operating conditions and service cycles as components, can serve as indicators of possible material degradation and susceptibility for the actual installation. They are evaluated upon their removal from the system. They are usually cleaned, weighed, and examined in the laboratory at periodic intervals, then replaced for further exposure. Monitoring can take place over extended periods and the results documented in detailed logs. This can be of significant assistance as reference data in avoiding material failures by corrosion or erosion-corrosion.[167–169]

In using results from these tests and in interpreting their response, it is necessary to recognize that their conditions may not completely duplicate conditions present in the actual component or vessel. That is, they may not corrode at the same rates or manner as in actual equipment where the metal is a heat-transfer medium. Other effects may also be introduced or absent that can have significant effects upon test outcomes and may not be representative of actual component response.

Other types of corrosion monitors have been devised. One is comprised of two metal elements of the desired composition to be evaluated. One is exposed to the environment while the other, the reference element, is protected. As the exposed element corrodes, its increased electrical resistance over that of the reference element is calculated in terms of corrosion rate in mils per year (mpy). This probe functions as an in situ sensor that accumulates the corrosion history of the environment it is exposed to and displays or records the corrosion information.[170] Once again, conditions at the surface of the corroding test element may not duplicate those of the actual component, and these possible differences should be taken into account when interpreting results.

Other types of corrosion monitoring devices operate on other principles, such as the relationship between a metal's corrosion potential and its corrosion state as a function of its corrosion rate (potential monitoring). Metering of thickness of scales, corrosion products, or base metal thicknesses also provide a basis for monitoring corrosion progress and traditional NDE methods have been employed in these evaluations.

Toughness monitoring. During the service life of many components, microstructural changes can occur in materials resulting in impaired toughness. This is caused by various embrittlement mechanisms, such as formation of molybdenum-rich metal carbides at the grain boundaries of chrome-molybdenum steels, temper embrittlement from phosphorus segregation, and sensitization effects in high-alloy steels. In heavy-section components of high restraint, severe thermal stresses from startup and shutdown can produce brittle fractures in such toughness-impaired materials. Knowledge of this condition can help to forestall failure.

In situ metallography can identify these conditions, but correlations between observed microstructures and quantitative toughness values is difficult. Toughness can be determined directly using Charpy V-notch test specimens removed from the actual component wall, but these destructive methods involve removing significant volumes of material for replicate tests over a range of temperatures. These sites would require weld repairs, and this could introduce other problems, such as additional residual stresses and questionable microstructures and compositional gradients. Accordingly, methods have been devised for monitoring toughness of such materials using small macrospecimens.[171]

For example, punch tests use a small wafer or disk of the material (0.25 in diameter × 0.020 in thickness or 6.35 mm diameter × 0.508 mm thickness) that has been removed from a selected surface of the component. Specially designed cutters minimize the volume of material required. The disk is inserted between a small hemispherical punch and its mating die. The wafer is deformed between the die members under a controlled rate of punch displacement, and load versus deflection curves are obtained. The area under the curve corresponds to absorbed energy. Tests conducted over a range of temperatures provide a temperature versus absorbed energy relationship for the material. Ductile–brittle transition temperatures derived for material in this test are reported to correlate well with those obtained in Charpy impact tests.

Fatigue monitoring. Because fatigue is identified as the fracture mechanism in most mechanical failures, methods for monitoring its initiation and progress have received a good deal of attention. It has been estimated from laboratory tests that a major portion of the fatigue life of a part (up to 90% or more) is spent during initiation (stage 1). This is particularly so for low-strain-rate amplitudes and ductile materials. It should be noted that the number of cycles to failure (N in S/N curves; see Fig. 3.11) represents the total number of cycles to failure and includes both initiation and propagation stages.

In view of the greater proportion of time required for crack initiation for many common engineering materials, it would seem to make sense to concentrate upon this stage during monitoring, as methods that could signal the end of stage 1 and the onset of stage 2 (crack propagation) would be most helpful in determining the lifetime of a component. Despite the amount of study that has been devoted to understanding the mechanics of stage 1, progress has been slow. To date, no reliable method is available for identifying the time of the stage 1 to stage 2 transition in operating components.

However, researchers can examine surfaces of fatigue test specimens in the laboratory as stage 1 begins and progresses to the onset of stage 2, and they have learned much about the sequence of microevents that take place during fatigue crack initiation. Crack initiation in smooth specimens is an extremely localized surface phenomenon that is apparently related to very localized surface regions where the stress level exceeds the elastic limit, resulting in localized plastic deformation (on a microscale).

Alternating localized plastic strains also create microirregularities at the surface, generally along slip planes in ductile materials. These constitute minute crevices surrounded by work-hardened or strain-hardened grains. Microintrusions can initiate at these locations and slowly progress inward, along active slip bands, at an inclined angle of about 45° to the direction of maximum principal stress. When this "stage 1" crack becomes large enough for the crack-tip stress field to dominate, its direction turns 45° or so and becomes normal to the direction of the applied tensile stress. At this stage (stage 2) the crack propagates until the load-carrying capacity of the specimen or part is sufficiently impaired to cause it to fail by mechanical overload (stage 3).[172]

For structural components containing surface defects, cracks, and flaws, crack initiation processes can occur in much less time. Stress-concentrating geometric features at the surface can have the same effect to foreshorten the initiation period. Surface cracks initiated by other mechanisms can proceed more or less directly into stage 2 propagation.

Accordingly, in the real world, the response of smoothly machined fatigue test specimens cycled under ideal laboratory testing conditions (e.g., zero mean stress, uniform amplitude, sinusoidal loading, ambient temperature and environment) may not be closely relevant in determining the fatigue life of an operating component. So, the time required for fatigue cracks to initiate in laboratory specimens may not be applicable elsewhere. Besides, actual stress cycles in dynamically loaded components rarely duplicate the constant-amplitude conditions typical of laboratory-conducted fatigue tests.

It is worth noting that although fatigue crack initiation may be accelerated in actual components by stress-concentrating geometries or preexisting flaws or cracks, propagation may be slower than in laboratory tests as a result of lower applied stress levels. This can mean that stage 2 fatigue cracks can exist for extended periods of time without being a threat to the life of the part.

Life-extension assessments of fatigue in operating equipment are focused upon crack propagation characteristics. As noted previously, this is dependent upon many factors, including applied nominal mean and alternating stress levels, the cycle profile, material properties, and the environment. Fatigue life of existing components, then, depends largely upon detection of existing cracks and capability for monitoring or evaluating their growth. Also required is information on fracture propagation properties of the material. This is important in establishing realistic inspection intervals so that maximum tolerable crack size can be determined. Since crack propagation in some materials can accelerate significantly in specific environments, these aspects should also be well defined.[173]

Maximum detectable flaw size for the material and NDE inspection method should be known. It has been reported that the smallest flaw that can be readily detected during routine service inspection is about 5 to 15 mm long (0.2 to 0.6 in).[174] Others have indicated values for various standard methods and these range from 3 to 5 mm (0.12 to 0.2 in) for aerospace components presumably inspected under ideal conditions (smooth, clean surfaces).[175] These values have limited significance unless the material thickness is specified along with the configuration of the region, as detectability is affected by these variables.

In determining the general state of a component with respect to fatigue life, an assessment of fracture toughness in terms of critical crack size should be related to the smallest detectable crack. A clear hazard exists if the critical crack size for the situation is smaller than or near to the lower detectable limit. Such a situation would require use of a more discriminating NDE procedure, if possible, or structural modification using tougher material.

Effective monitoring of fatigue cracking, therefore, requires more than periodic measurements of crack lengths or the detection of propagating fatigue cracks. Growth rates determine inspection intervals, and usefulness of crack monitoring in the first place is directly related to critical flaw sizes and detectability limits for applicable NDE techniques.

Traditional fatigue crack monitoring was (and still is) based upon S/N behavior of the material. Counters monitor number of loading cycles and this is used to estimate damage accumulation in evaluating remaining component life. The levels of mean and alternating stresses govern the number of cycles to failure; therefore, good understanding of this is necessary in applying these concepts.

Design load values or calculated operating stresses may not be appropriate or correct for these evaluations, as localized stresses in various regions can be higher, as can residual stress from welding, distortion, shrink-fit parts, and dissimilar joints. More accurate information concerning stress levels can be obtained through strain-gage measurements or other assessment techniques that have been described. However, their applicability to operating equipment can be limited, as at least some of the stress applied to fatigue cracks may be inherent in the existing structure or component. Strain gages will show changes that occurred from the time of their attachment or installation.

The application of fatigue monitoring methodologies to existing equipment in evaluating fatigue life is confronted with many unknowns that can influence their effectiveness. A major question at the outset is the amount of accumulated damage that has already been sustained. NDE techniques can ascertain if cracks are present and their sizes. Operating logs, equipment instrumentation, and historical information can provide estimates of loading history and number of stress cycles. And information on the material, its properties, and service background can indicate basic expected response.

As formidable as the situation may seem, an orderly and conscientious search for relevant information, coupled with competent NDE and a monitoring program providing operating cycle and crack-growth information, can be effective in forestalling fatigue failures. A combined approach that integrates damage mechanics estimates based upon S/N data analysis and fracture mechanics–based crack-growth rate analysis probably has the best chance of success.[176] However, to date, most programs are based upon concepts of the first type (damage mechanics estimates).

For example, a lifetime prediction method that has been evaluated for power plants involves automated monitoring technology that utilizes plant process data to perform a continuous prediction of fatigue damage accumulation in critical components.[177] The PC-based program acquires plant process computer data (e.g., temperatures, pressures, flows, etc.) and interprets them to predict local loads and temperatures in monitored components. Stresses are then calculated and a fatigue evaluation is made based upon computerized stress history. A fatigue analysis based upon linear cumulative damage estimates (Miner's rule) is conducted in accordance with ASME Boiler and Pressure Vessel Code procedures.[178] This is an example of application of damage mechanics estimates. Examples of fatigue lifetime prediction on a fracture mechanics basis are described in the literature.[179–181]

Crack-growth-rate characteristics for an operating component are important in estimating fatigue life. NDE methods can identify cracks within the limits of detectability of inspection equipment, and these inspections are usually carried out during scheduled shutdowns. Cracks of less than critical size will possibly increase in size during continued operation. Key questions are: how fast will they grow and how often should they be monitored? Crack-growth curves of the type shown in Fig. 3.12 have been published for various commercial alloys and may be used to estimate growth rates of existing cracks by calculating K_I values from NDE crack size information.

Crack-progress monitoring devices have been developed around standard NDE principles, with ultrasonic methods being predominant because of greater inherent reliability and detectability.[182] The development of proprietary devices reported recently involves affixing sensors ("fatigue fuses") to critical regions of structures and components. The sensors are designed with various predictable fatigue lives of their own, and a graded array of these is affixed to regions of interest on the component or structure. These sensors respond to fatigue loading of the component and offer indications of accumulated damage sustained when signals of successive failures of individual sensors are relayed back to a centralized location through microprocessor-controlled equipment.[183]

Other approaches are also being studied. Although most are not yet sufficiently developed to a stage amenable to use in actual crack-growth sensors, a number of concepts are being pursued for on-line monitoring of operating equipment. Techniques are based upon acoustic, electrical, thermal, x-ray, neutron, magnetic, and optical methods.[184] Not all are focused upon cracks, per se, but are also evaluating mechanical and microstructural properties of materials affected by propagating cracks and other flaws.[185–188]

Monitor integrity. Throughout this section, we have discussed use of various types of condition and material assessment monitors. These can play vital roles in a life-extension program. Some of the monitors are quite simple while others are complex. Since their response and indications constitute basic input to these programs and activities, their reliability is critical and deserves attention.[189] A number of serious failures have been attributed to malfunctioning monitors and to such simple incidents as burned-out indicator bulbs and false indications, as demonstrated by the following example.

In an earlier chapter we noted how failures are precipitated by a broad array of causal elements, many unforeseen. The incident in 1979 at the Three Mile Island nuclear power plant was no exception. Although the extent of the damage was reportedly compounded by a series of inappropriate responses to control room indications, an identified major contributor was a spring-actuated-return relief valve in the pressurized primary water line that had stuck in the open position.

Unfortunately, the light on the control panel indicating the valve was closed merely reported that the solenoid had been signaled to close the valve. It offered no assurance that the valve was actually in a closed position, as the indicator light was wired to respond to solenoid activation commands, not to relief valve status, that is, there was no provision to compensate for the possibility of valve return spring malfunction or failure or for alerting operators to such a condition. Despite symptoms that things were not as they should have been due to insufficient coolant in the reactor, plant operators chose to rely upon the valve "position" indicator. This prompted them to shut down water pumps, compounding the problem.[190–192]

Aside from using duplicate monitors, which may or may not improve reliability, monitors might be instruments with self-test features, as is common practice today in electronic systems. Operational status could be indicated automatically or the monitors could be wired to permit manual checks of operational status. For critical systems, an automatic self-test feature with builtin alert signals should be provided.

Increasing need for monitoring in manufacturing systems and process plants has prompted development of improved instruments operating by advanced technologies along with more environmentally resistant packaging. Operating lifetimes of monitors themselves and their survival and reliability under required conditions should be well established before using any monitor for critical service applications.[193–195]

Destructive testing. Discussions so far on assessing the condition of materials in estimating remaining life were confined mostly to nondestructive methods, although some techniques required removal of surface scale and irregularities and other relatively superficial sampling of material. A few destructive-type tests were mentioned briefly in passing whenever they were relevant.

Destructive testing does not imply that the component's life is over. It may be, if tests are conducted in conjunction with failure analysis, as is often the case. However, in the context of the present subject of monitoring in assessing material condition as it relates to remaining service life, destructive testing refers to sampling of material from actual operating systems and components. This is done to provide material for laboratory tests to characterize more completely the condition of this equipment than is possible by other methods.

A great deal of reliable information can be obtained from relatively small samples removed from operating equipment, provided the tests are carefully planned and executed.[196] Usually, this technique is used only when nondestructive methods cannot provide desired information, when they produce inconclusive results, or when an important decision requires the best available data on material condition. Sample removal can usually be scheduled only during shutdown periods when locations of interest are accessible and when sampling operations cannot disrupt the equipment or plant output.

Sample removal and retrieval represent but one step of the process, as satisfactory repairs must be made of sampled regions before system or plant operation can be resumed. Also, the repairs must usually be made in accordance with applicable fabrication standards governing materials, procedures, welder competence, and post-repair inspection and acceptance standards. It can be a costly and time-consuming operation. Both before and after material sampling, photographs showing sample locations on various scales (i.e., camera-to-object distances) should be taken, with clear and unambiguous identifications and records to verify sample location, size, its orientation, how it was removed, repairs made, and sample disposition.

Sampling environments are often adverse. Dirt, dust, and other hazards call for adequate personnel protection and safety practices. Nevertheless, precautions should be taken to avoid contaminating the sample or damaging it. These include protection from cutting lubricants, torch spatter, grinding dusts, and sawing debris. Handling should be minimized and protective gloves should be worn to avoid physical contact with the sample and its possible contamination. In the author's experience, wasted boiler tube samples that had been removed for metallurgical examination and analysis were received well wrapped, protected, and sealed in containers with dessicants. However, adhesive plastic identifying labels had been firmly affixed to the surfaces of interest.

It is worthwhile to devise tests that require minimal volumes of material, as this allows more tests to be run, perhaps over a range of temperatures, or for additional types of tests to be conducted. Test sequence should be planned so that surface conditions and effects are evaluated first. This will avoid damaging these surfaces during sectioning or other test preparation steps. The tests that may be required will depend upon the application and purpose for removing the samples. Table 2.1 lists some of the tests and procedures that may be required during failure investigations, and these may also be carried out, as necessary, on samples removed from operating equipment.

Can We Afford It?

As difficult as it may be to estimate the remaining life of operating components, an even more difficult task is to determine how much money should be spent on these programs. To further complicate matters, it is not a one-time expenditure but an ongoing activity. There is much more to it than purchase and installation of NDE or monitoring equipment, as life-extension analyses require expertise that may not exist within the organization. It may require data collection, storage and retrieval systems, computer software and perhaps its development, laboratory testing programs and specialized diagnostic equipment, and other things.

How extensive the system should be depends upon the application, size and complexity of the operation involved, and how critical it is to extend its useful life and to avoid unplanned outages and failures. The major costs of the program probably do not lie within the monitoring activity itself, but in implementing indicated recommendations and in performing the various hardware and operational modifications that are required to achieve desired goals.

These are major considerations and decisions affecting depth and extent of the program. Adoption of these strategies and the degree to which they are implemented require a comprehensive and competent cost-effectiveness analysis to supplement personal intuition and judgment.[197] There is no practical alternative as a basis for these decisions because we are dealing with a significant financial commitment that will continue for the life of the plant or venture. Unless there is a sound, logical, and defensible basis for the expenditure and commitment, the activity will itself fail. A formalized approach provides a tangible result that can be evaluated objectively and documented, and this—of itself—can provide a significant advantage.

Cost decisions made for implementing programs in one plant or facility cannot be reasonably extrapolated to another. Each one must be evaluated on its own. A broadly generalized decision model was described earlier in this chapter for implementing failure avoidance strategies, and costs should have been one of the principal considerations in the initial decision. Presuming that this or a similar decision model was applied in analyzing a given situation and that it has been decided that a failure avoidance program is required, similar decisions must be made for budget allocation as this determines extent of the program and what can be done.

Major considerations at this stage include consequences of system failures in the absence of implemented avoidance strategies or under an inadequate program. Another major consideration involves the need for extending usable life of the system or equipment. Perhaps it is the only available option to closing down the operation. Capitalization for total replacement of an aging system may be unavailable, or political or environmental considerations may prohibit new construction; many other factors can dictate the direction of these decisions and financial resources that can be committed.

On the other hand, perhaps the aging system involves obsolete or obsolescing technology and should be discontinued. Perhaps basic elements can be retained and upgraded for use within a new facility based upon new technology. The needs can differ widely. However, the strategies that have been described constitute another option to plant write-off and operational termination, that is, provided that the operation is essential and profitable, that its technology is current or adequately upgradable, that the system has not yet reached the end of its useful life, and that life-extension methodologies will prove cost effective for the given installation.

For facilities constructed as recently as 20 years ago, their designs probably were not based upon the newer technologies of fracture mechanics and damage-tolerant practices. Since then, understanding of failure mechanisms and materials science has increased considerably. It is often found, in the light of present knowledge in these respects, that existing systems engineered for 25- to 35-year life by traditional methods were designed on the basis of overly conservative assumptions. As a result, many facilities can safely operate for some time longer provided appropriate life-extension assessment and failure avoidance strategies are implemented.

References

1. C. West Churchman, *The Systems Approach,* Dell, New York, 1979.
2. B. H. Rudwick, *Systems Analysis for Effective Planning—Principles and Cases,* Wiley, New York, 1969.
3. D. I. Cleland and W. R. King, *Systems Analysis and Project Management,* McGraw-Hill, New York, 1968.
4. R. A. Johnson, F. E. Kast, and J. E. Rosenzweig, *The Theory and Management of Systems,* McGraw-Hill, New York, 1967.
5. Simon Ramo, *Century of Mismatch,* David McKay, New York, 1970, pp. 143–144.
6. Simon Ramo, *Cure for Chaos: Fresh Solutions to Social Problems Through the Systems Approach,* David McKay, New York, 1969.
7. J. Morley English (Ed.), *Cost-Effectiveness, The Economic Evaluation of Engineered Systems,* Wiley, New York, 1968, pp. 11–32, 113–165, 214–241.
8. Harris R. Greenberg and Joseph J. Cramer, *Risk Assessment and Risk Management for the Chemical Process Industry,* Van Nostrand Reinhold, New York, 1991, pp. 15–29, 48–56, 101–126.
9. Willie Hammer, *Handbook of System and Product Safety,* Prentice-Hall, Englewood Cliffs, NJ, 1972, pp. 62–140, 322–342.
10. Floyd R. Tuler, "Risk Assessment—Offshore Oil and Gas Operations," *Mechanical Engineering,* Vol. 106, No. 11, November 1984, pp. 24–30.
11. S. B. Gibson, "Hazard Analysis and Risk Criteria," *Loss Prevention,* Vol. 14, American Institute of Chemical Engineers, 1980, pp. 11–17.
12. Ernest J. Henley and Hiromitsu Kumamoto, *Reliability Engineering and Risk Assessment*, Prentice-Hall, Englewood Cliffs, NJ, 1981, pp. 8–43.
13. Harris R. Greenberg and Joseph J. Cramer, Ref. 8, pp. 30–47.
14. John Kolb and Steven S. Ross, *Product Safety and Liability—A Desk Reference,* McGraw-Hill, New York, 1980, pp. 99–100, 329–377.
15. H. E. Lambert, *Fault Trees for Decision Making in Systems Analysis,* Lawrence Livermore National Laboratory, Livermore, CA, Report No. UCRL-51829, October 9, 1975.
16. Willie Hammer, Ref. 9, pp. 113–115.
17. Ernest J. Henley and Hiromitsu Kumamoto, Ref. 12, pp. 44–108.
18. Harris R. Greenberg and Joseph J. Cramer, Ref. 8, pp. 127–166.
19. Willie Hammer, Ref. 9, pp. 238–246.
20. John Kolb and Steven S. Ross, Ref. 14, pp. 115–129.
21. J. S. Arendt and J. B. Fussell, "System Reliability Engineering Methodology for Industrial Application," *Loss Prevention,* Vol. 14, American Society of Chemical Engineers, 1980, pp. 18–28.
22. William E. Vesely, "Failure Data and Risk Analysis," *Failure Data and Failure Analysis: In Power and Processing Industries,* PVP-PB-023, The American Society of Mechanical Engineers, New York, 1977, pp. 61–75.
23. Steven A. Lapp and Gary J. Powers, "A Method for the Generation of Fault Trees," *Failure Data and Failure Analysis: In Power and Processing Industries,* PVP-PB-023, The American Society of Mechanical Engineers, New York, 1977, pp. 95–101.
24. Richard E. Barlow, Jerry B. Fussell, and Nozer D. Singpurwalla, *Reliability and Fault Tree Analysis—Theoretical and Applied Aspects of System Reliability and Safety Assessment,* Society for Industrial and Applied Mathematics, Philadelphia, PA, 1975.
25. H. E. Lambert, Ref. 15.
26. R. W. Prugh, "Application of Fault Tree Analysis, *Loss Prevention,* Vol. 14, American Institute of Chemical Engineers, 1980, p. 1.
27. Ernest J. Henley and Hiromitsu Kumamoto, Ref. 12, pp. 288–361, xv–xix.
28. Harris R. Greenberg and Joseph J. Cramer, Ref. 8, pp. 91–100.
29. Willie Hammer, Ref. 9, pp. 148–156.
30. John Kolb and Steven S. Ross, Ref. 14, pp. 129–147.
31. *Design Analysis Procedure for Failure Modes, Effects and Criticality Analysis (FMECA)*, SAE Aerospace Recommended Practice, ARP926, 15 September 1967.

32. K. Greene, "Failure Mode, Effects and Criticality Analysis," *PLP-79 Proceedings,* Product Liability Prevention Conference, New York, October, 1979, IEEE Catalog No. 79CH1512-3R, p. 55.
33. *U.S. MIL-STD-1629,* "Failure Mode and Effects Analysis," National Technical Information Service (NTIS), Springfield, VA.
34. A. B. Mundel, "Failure Modes and Effects Analysis as a Means of Product Liability Prevention," *Proceedings PLP/75,* Product Liability Prevention Conference, Newark, NJ, August, 1975, p. 61.
35. J. I. Dickson, E. Abramovici, and N. S. Marchand (Eds.), *Failure Analysis: Techniques and Application,* Proceedings of the First International Conference on Failure Analysis, ASM International, Materials Park, OH, July, 1991, pp. 263–318.
36. William S. Pellini, *Principles of Structural Integrity Technology,* Office of Naval Research, Arlington, VA, 1976, pp. 87–124.
37. William S. Pellini, "Principles of Fracture-Safe Design," *Welding Journal,* Research Supplement, Vol. 50, Nos. 3–4, 1971, pp. 91s–108s, 147s–162s.
38. "Standard Test Methods for Notched Bar Impact Testing of Metallic Materials," *ASTM E23, Annual Book of ASTM Standards,* American Society for Testing and Materials, Philadelphia, PA, 1993.
39. "Standard Test Method for Dynamic Tear Testing of Metallic Materials," *ASTM E604, Annual Book of ASTM Standards,* American Society for Testing and Materials, Philadelphia, PA, 1993.
40. "Standard Method for Drop Weight Tear Tests of Ferritic Steels," *ASTM E436, Annual Book of ASTM Standards,* American Society for Testing and Materials, Philadelphia, PA, 1993.
41. "Standard Method of Sharp-Notch Tension Testing of High-Strength Sheet Materials," *ASTM E338, Annual Book of ASTM Standards,* American Society for Testing and Materials, Philadelphia, PA, 1993.
42. "Standard Method for Sharp-Notch Tension Testing with Cylindrical Specimens," *ASTM E602, Annual Book of ASTM Standards,* American Society for Testing and Materials, Philadelphia, PA, 1993.
43. "Standard Practice for Fracture Testing with Surface-Crack Tension Specimens," *ASTM E740, Annual Book of ASTM Standards,* American Society for Testing and Materials, Philadelphia, PA, 1993.
44. Stanley T. Rolfe and John M. Barsom, *Fracture and Fatigue Control in Structures—Applications of Fracture Mechanics,* Prentice-Hall, Englewood Cliffs, NJ, 1977.
45. J. F. Knott, *Fundamentals of Fracture Mechanics,* Butterworths, London, 1973.
46. "Standard Test Method for Plane-Strain Fracture Toughness of Metallic Materials," *ASTM E399, Annual Book of ASTM Standards,* American Society for Testing and Materials, Philadelphia, PA, 1993.
47. "Standard Terminology Relating to Fracture Testing," *ASTM E616, Annual Book of ASTM Standards,* American Society for Testing and Materials, Philadelphia, PA, 1993.
48. "Standard Practices for Plain-Strain Fracture Toughness Testing of Aluminum Alloys," *ASTM B645, Annual Book of ASTM Standards,* American Society for Testing and Materials, Philadelphia, PA, 1993.
49. "Standard Practices for Fracture Toughness Testing of Aluminum Alloys," *ASTM B646, Annual Book of ASTM Standards,* American Society for Testing and Materials, Philadelphia, PA, 1993.
50. "Standard Test Method for Determining Plane-Strain Crack-Arrest Fracture Toughness, K_{Ia} of Ferritic Steels," *ASTM E1221, Annual Book of ASTM Standards,* American Society for Testing and Materials, Philadelphia, PA, 1993.
51. "Standard Test Method for J_{IC}, A Measure of Fracture Toughness," *ASTM E813, Annual Book of ASTM Standards,* American Society for Testing and Materials, Philadelphia, PA, 1993.
52. "Standard Practice for *R*-Curve Determinations," *ASTM E561, Annual Book of ASTM Standards,* American Society for Testing and Materials, Philadelphia, PA, 1993.
53. "Standard Test Method for Crack-Tip Opening Displacement (CTOD) Fracture

Toughness Measurement," *ASTM E1290, Annual Book of ASTM Standards,* American Society for Testing and Materials, Philadelphia, PA, 1993.
54. Stanley T. Rolfe and John M. Barsom, Ref. 44, pp. 167–207.
55. John E. Srawley and Jack B. Esgar, "Investigation of Hydrotest Failure of Thiokol Chemical Corporation 260-Inch-Diameter SL-1 Motor Case," *NASA TMX-1194,* National Aeronautics and Space Administration, Lewis Research Center, Cleveland, OH, January, 1966, U.S. Department of Commerce, NTIS, N66-14708.
56. Kenneth Easterling, *Introduction to the Physical Metallurgy of Welding,* Butterworths, London, 1983, pp. 203–220.
57. B. F. Brown, *Stress Corrosion Cracking Control Measures,* National Association of Corrosion Engineers, Houston, TX, 1981, pp. 1–2.
58. W. Rostoker, J. M. McCaughey, and H. Markus, *Embrittlement by Liquid Metals,* Reinhold, New York, NY, 1960.
59. Mars G. Fontana and Norbert D. Greene, *Corrosion Engineering,* 2nd ed., McGraw-Hill, New York, NY, 1978, pp. 91–107.
60. *Metals Handbook,* 9th ed., Vol. 11, *Failure Analysis and Prevention,* American Society for Metals, Metals Park, OH, 1986, pp. 203–224.
61. D. V. Beggs and M. T. Hahn, "Recent Observations on the Propagation of Stress Corrosion Cracks and Their Relevance to Proposed Mechanisms of Stress Corrosion Cracking," *Hydrogen Embrittlement and Stress Corrosion Cracking* (R. Gibala and R. F. Hehemann, Eds.), American Society for Metals, Metals Park, OH, 1984, pp. 181–316.
62. *Metals Handbook,* Ref. 60, p. 203.
63. Ibid., p. 206.
64. Herbert H. Uhlig, *Corrosion and Corrosion Control,* 2d ed., Wiley, New York, 1971, p. 130.
65. "Standard Test Methods for Chemical Analysis of Thermal Insulation Materials for Leachable Chloride/Fluoride/Silicate/Sodium Ions," *ASTM C871, Annual Book of ASTM Standards,* American Society for Testing and Materials, Philadelphia, PA, 1993.
66. *Metals Handbook,* Ref. 60, p. 207.
67. B. F. Brown, Ref. 57.
68. D. H. Thompson, *Stress Corrosion Cracking of Metals—A State of the Art* (H. Lee Craig, Jr., Ed.), STP-518, American Society for Testing and Materials, Philadelphia, PA, 1972.
69. L. L. Shreir, *Corrosion,* Vol. 1, Wiley, New York, 1963, pp. 8.3–8.70.
70. Markus O. Speidel, "Stress Corrosion Cracking of Aluminum Alloys," *Metallurgical Transactions A,* Vol. 6A, April, 1975, pp. 631–651.
71. "Standard Practice for Making and Using U-Bend Stress-Corrosion Test Specimens," *ASTM G30, Annual Book of ASTM Standards,* American Society for Testing and Materials, Philadelphia, PA, 1993.
72. "Standard Practice for Evaluating Stress-Corrosion Cracking Resistance of Metals and Alloys in a Boiling Magnesium Chloride Solution," *ASTM G36, Annual Book of ASTM Standards,* American Society for Testing and Materials, Philadelphia, PA, 1993.
73. "Standard Practice for Making and Using C-Ring Stress-Corrosion Test Specimens," *ASTM G38, Annual Book of ASTM Standards,* American Society for Testing and Materials, Philadelphia, PA, 1993.
74. "Standard Practice for Preparation and Use of Bent-Beam Stress-Corrosion Test Specimens," *ASTM G39, Annual Book of ASTM Standards,* American Society for Testing and Materials, Philadelphia, PA, 1993.
75. "Standard Practice for Evaluating Stress Corrosion Cracking Resistance of Metals and Alloys by Alternate Immersion in 3.5% Sodium Chloride Solution," *ASTM G44, Annual Book of ASTM Standards,* American Society for Testing and Materials, Philadelphia, PA, 1993.
76. "Standard Test Method for Determining Susceptibility to Stress-Corrosion Cracking of High-Strength Aluminum Alloy Products," *ASTM G47, Annual Book of ASTM Standards,* American Society for Testing and Materials, Philadelphia, PA, 1993.

77. "Standard Practice for Preparation and Use of Direct Tension Stress-Corrosion Test Specimens," *ASTM G49, Annual Book of ASTM Standards,* American Society for Testing and Materials, Philadelphia, PA, 1993.
78. "Standard Practice for Preparation of Stress-Corrosion Test Specimens for Weldments," *ASTM G58, Annual Book of ASTM Standards,* American Society for Testing and Materials, Philadelphia, PA, 1993.
79. Stanley T. Rolfe and John M. Barsom, Ref. 44, p. 292.
80. Ibid., pp. 301–313.
81. B. F. Brown, Ref. 57.
82. A. R. Troiano, "The Role of Hydrogen and Other Interstitials in the Mechanical Behavior of Metals," *Transactions of the American Society for Metals,* Vol. 50, 1960, pp. 54–80.
83. H. R. Gray, *Hydrogen Embrittlement Testing, ASTM STP-543,* American Society for Testing and Materials, Philadelphia, PA, 1974, pp. 3–5.
84. J. P. Fidelle and H. G. Nelson, "Closing Commentary—IHE-HEE: Are They the Same?" *Hydrogen Embrittlement Testing, ASTM STP-543,* American Society for Testing and Materials, Philadelphia, PA, 1974, pp. 267–274.
85. C. D. Kim, "Hydrogen-Damage Failures," *Metals Handbook,* 9th ed., Vol. 11, American Society for Metals, Metals Park, OH, 1986, p. 250.
86. H. K. Birnbaum, "Hydrogen-Related Second Phase Embrittlement of Solids," *Hydrogen Embrittlement and Stress Corrosion Cracking* (R. Gibala and R. F. Hehemann, Eds.), American Society for Metals, Metals Park, OH, 1984, pp. 153–177.
87. J. C. Williams, "Hydride Formation," in *Effect of Hydrogen on Behavior of Materials* (A. W. Thompson and I. M. Bernstein, Eds.), The Metallurgical Society of AIME, New York, NY, 1976, pp. 367–380.
88. Paul G. Shewmon, "Hydrogen Attack of Carbon Steel," *Effect of Hydrogen on Behavior of Materials* (A. W. Thompson and I. M. Bernstein, Eds.), The Metallurgical Society of AIME, New York, 1976, pp. 59–69.
89. S. W. Ciaraldi, "Selection of Petroleum Industry Material Through Use of Environmental Cracking Tests," *Hydrogen Embrittlement: Prevention and Control, ASTM STP-962* (L. Raymond, Ed.), American Society for Testing and Materials, Philadelphia, PA, 1988, pp. 200–213.
90. T. P. Groeneveld and A. R. Elsea, "Mechanical Testing Methods," Ref. 83, p. 11.
91. Herbert A. Johnson, "Keynote Lecture: Overview on Hydrogen Degradation Phenomena," Ref. 61, p. 9.
92. John J. DeLuccia, "Electrochemical Aspects of Hydrogen in Metals," Ref. 89, pp. 17–34.
93. C. D. Kim, Ref. 85, p. 246.
94. Edward T. Clegg, "Hydrogen Embrittlement Coverage by U.S. Government Standardization Documents," Ref. 89, pp. 37–45.
95. H. P. Gray, "Testing for Hydrogen Embrittlement Experimental Variables," Ref. 83, pp. 133–151.
96. J. P. Fidelle, R. Broudeur, C. Pirrovani, and C. Roux, "Disk Pressure Technique," Ref. 83, pp. 34–47.
97. J. P. Fidelle, R. Bernardi, R. Broudeur, C. Roux, and M. Rapin, "Disk Pressure Testing of Hydrogen Environment Embrittlement," Ref. 83, pp. 221–253.
98. Edward T. Clegg, Ref. 94.
99. Stephen D. Antolovich and Ashok Saxena, "Fatigue Failures," *Metals Handbook,* Vol. 11, Failure Analysis and Prevention, American Society for Metals, Metals Park, OH, 1986, p. 135.
100. Reinhold H. Dauskardt, Robert O. Ritchie, and Brian N. Cox, "Fatigue of Advanced Materials: Part I," *Advanced Materials and Processes,* July, 1993, p. 26.
101. N. E. Frost, K. J. Marsh, and L. P. Pook, *Metal Fatigue,* Clarendon, Oxford, 1974.
102. Carl C. Osgood, *Fatigue Design,* Wiley, New York, 1970, p. 365.
103. "Standard Practice for Conducting Constant Amplitude Axial Fatigue Tests of Metallic Materials," *ASTM E466, Annual Book of ASTM Standards,* American Society for Testing and Materials, Philadelphia, PA, 1993.

104. "Standard Test Method for Measurement of Fatigue Crack Growth Rates," *ASTM E647, Annual Book of ASTM Standards,* American Society for Testing and Materials, Philadelphia, PA, 1993.
105. R. P. Skelton (Ed.), *Fatigue At High Temperature,* Applied Science, Essex, England, 1983.
106. Hermann Riedel, *Fracture At High Temperature,* Springer, New York, 1987, pp. 51–260.
107. Stephen D. Antolovich and Ashok Saxena, Ref. 99, p. 133.
108. R. Hales, "Fatigue Testing Methods at Elevated Temperatures," *Fatigue At High Temperatures* (R. P. Skelton, Ed.), Applied Science, Essex, England, 1983, pp. 63–67.
109. O. F. Devereux, A. J. McEvely, and R. W. Staehle (Eds.), *Corrosion Fatigue: Chemistry, Mechanics and Microstructure,* National Association of Corrosion Engineers, Houston, TX, 1972.
110. Henry J. Cialone and John Holbrook, "Sensitivity of Steels to Degradation in Gaseous Hydrogen," *Hydrogen Embrittlement: Prevention and Control, ASTM STP-962* (L. Raymond, Ed.), American Society for Testing and Materials, Philadelphia, PA, 1988, pp. 134–152.
111. Ibid., p. 149.
112. Warren F. Savage, "Solidification, Segregation, and Weld Defects," *Weldments: Physical Metallurgy and Failure Phenomena* (R. J. Christoffel, E. F. Nippes, and H. D. Solomon, Eds.), General Electric Co., Schenectady, NY, 1979, pp. 1–18.
113. *Metals Handbook,* 9th ed., Vol. 6, "Welding, Brazing, and Soldering," American Society for Metals, Metals Park, OH, 1983.
114. *Welding Handbook,* 8th ed. (currently in Vol. 2 of 3 volumes), American Welding Society, Miami, FL.
115. George E. Linnert, *Welding Metallurgy—Carbon and Alloy Steels,* 3d ed., 2 volumes, American Welding Society, Miami, FL, 1967.
116. T. G. F. Gray, J. Spence, and T. H. North, *Rational Welding Design,* Newnes-Butterworths, London, 1975.
117. Robert D. Stout and W. D. Doty, *Weldability of Steels,* 3d ed., Welding Research Council, New York, 1978.
118. Omer W. Blodgett, *Design of Welded Structures,* The James F. Lincoln Foundation, Cleveland, OH.
119. Kenneth Easterling, Ref. 56.
120. Howard B. Cary, *Modern Welding Technology,* Prentice-Hall, Englewood Cliffs, NJ, 1989.
121. *Design of Weldments,* The James F. Lincoln Foundation, Cleveland, OH.
122. Koichi Masubuchi, *Analysis of Welded Structures—Residual Stresses, Distortion and Their Consequences,* Pergamon, New York, 1980.
123. R. Viswanathan, *Damage Mechanisms and Life Assessment of High-Temperature Components,* ASM International, Materials Park, OH, 1989, pp. 1–2.
124. A. Hall, "The Bhopal Tragedy Has Union Carbide Reeling," *Business Week,* December 17, 1984, p. 32.
125. A. J. Parisi, "The Hard Lesson of Bhopal: It Can Happen Again," *Business Week,* December 24, 1984, p. 61.
126. Trevor Kletz, *Learning From Accidents in Industry,* Butterworths, London, 1988, pp. 83–91.
127. R. Viswanathan and S. M. Gehl, "Life Assessment Technology for Power Plant Components," *Journal of Metals,* Vol. 44, No. 2, February, 1992, pp. 34–42.
128. V. V. Bolotin, *Prediction of Service Life for Machines and Structures,* ASME, New York, 1989, pp. 17–19.
129. B. J. Cane and J. A. Williams, "Remaining Life Prediction of High Temperature Materials," *International Materials Reviews,* Vol. 32, No. 5, 1987, pp. 241–262.
130. *Metals Handbook,* 9th ed., Vol. 17, Non-Destructive Evaluation and Quality Control, ASM International, Metals Park, OH, 1989.
131. "Standard Practice for Liquid Penetrant Inspection Method," *ASTM E165, Annual Book of ASTM Standards,* American Society for Testing and Materials, Philadelphia, PA, 1993.

132. "Standard Practice for Magnetic Particle Examination," *ASTM E709, Annual Book of ASTM Standards,* American Society for Testing and Materials, Philadelphia, PA, 1993.
133. "Standard Practice for Standardizing Equipment for Electromagnetic Examination of Seamless Aluminum-Alloy Tube," *ASTM E215, Annual Book of ASTM Standards,* American Society for Testing and Materials, Philadelphia, PA, 1993.
134. "Standard Practice for Electromagnetic (Eddy Current) Testing of Seamless and Welded Tubular Products, Austenitic Stainless Steel and Similar Alloys," *ASTM E426, Annual Book of ASTM Standards,* American Society for Testing and Materials, Philadelphia, PA, 1993.
135. "Standard Guide for Radiographic Testing," *ASTM E94, Annual Book of ASTM Standards,* American Society for Testing and Materials, Philadelphia, PA, 1993.
136. "Standard Practice for Ultrasonic Pulse-Echo Straight-Beam Evaluation by the Contact Method," *ASTM E114, Annual Book of ASTM Standards,* American Society for Testing and Materials, Philadelphia, PA, 1993.
137. *Metals Handbook,* Ref. 130, pp. 278–294.
138. "Standard Practice for Acoustic Emission Monitoring of Structures During Controlled Stimulation," *ASTM E569, Annual Book of ASTM Standards,* American Society for Testing and Materials, Philadelphia, PA, 1993.
139. Stanley T. Rolfe and John M. Barsom, Ref. 44, pp. 508–510.
140. *Metals Handbook,* Ref. 130, pp. 663–715.
141. Barbara L. Gabriel, *SEM: A User's Manual for Materials Science,* American Society for Metals, Metals Park, OH, 1985, pp. 44–49.
142. Joel Fagenbaum, "Nondestructive Evaluation," *Mechanical Engineering,* May 1982, pp. 28–40.
143. Anon., "A High-Speed Chase," *Mechanical Engineering,* March, 1988, pp. 60–62.
144. R. C. McMaster, Ed., "Leak Testing," *Nondestructive Testing Handbook,* Vol. 1, 2d ed., American Society for Nondestructive Testing, Columbus, OH, 1982.
145. Gerald L. Anderson, "Leak Testing," *Metals Handbook,* 9th ed., Vol. 17, ASM International, Metals Park, OH, 1989, pp. 57–70.
146. John S. Mitchell, "Condition Monitoring," *Mechanical Engineering,* December, 1985, pp. 32–37.
147. Stomatios N. Thanos, "No Disassembly Required," *Mechanical Engineering,* September, 1987, pp. 86–90.
148. Jerome Rosen, "Power Plant Diagnostics Go On-Line," *Mechanical Engineering,* December, 1989, pp. 38–42.
149. B. J. Cane and J. A. Williams, Ref. 129, p. 256.
150. R. Viswanathan and S. M. Gehl, Ref. 127, op. cit.
151. B. J. Cane and J. A. Williams, Ref. 129, p. 255.
152. L. D. Lineback, "Strain Measurement for Stress Analysis," *Metals Handbook,* 9th ed., Vol. 17, American Society for Metals, Metals Park, OH, 1989, p. 453.
153. H. J. Macke and T. D. Sant, "The Intricate Patterns of Stress," *Mechanical Engineering,* December, 1982, pp. 19–23.
154. Koichi Masubuchi, Ref. 122, pp. 112–147.
155. J. F. Henry, F. V. Ellis, and R. Viswanathan, "Field Metallography Techniques for Plant Life Extension," *Microstructural Science,* Vol. 15 (M. E. Blum, P. M. French, R. M. Middleton, and G. F. Vander Voort, Eds.), 1987, pp. 13–26.
156. James J. Balaschak and Bernard M. Strauss, "Field Metallography in Assessment of Steam Piping in Older Fossil Power Plants," Ref. 155, pp. 27–36.
157. A. Cervoni and M. A. Clark, "Investigation of Turbine Disk Cracking by Field Metallography," Ref. 155, pp. 37–53.
158. S. M. Bruemmer, "Monitoring and Detecting Environmentally-Induced Cracking," *Journal of Metals,* December, 1990, p. 16.
159. I. M. Bernstein and A. W. Thompson, "The Role of Microstructure in Hydrogen Embrittlement, *Hydrogen Embrittlement and Stress Corrosion Cracking* (R. Gibala and R. F. Hehemann, Eds.), American Society for Metals, Metals Park, OH, 1984, p. 135.

160. N. Flanders, R. Tennant, and W. E. White, "Observations on Relationships Between Microstructure and Hydrogen-Induced Cracking," Ref. 155, pp. 227–240.
161. G. Calavre, J. C. Fontaine, A. Galibois, and M. R. Krishnadev, "Hydrogen Embrittlement Studies of Pearlitic Microstructures," Ref. 155, pp. 241–250.
162. I. M. Bernstein and A. W. Thompson, *International Metals Reviews,* Vol. 21, 1976, pp. 269–297.
163. L. Hagn, "Life Prediction Methods for Aqueous Environments," *Materials Science and Engineering, A103,* 1988, pp. 193–205.
164. Peter L. Andresen and F. Peter Ford, "Life Prediction by Mechanistic Modeling and System Monitoring of Environmental Cracking of Iron and Nickel Alloys in Aqueous Systems," *Materials Science and Engineering, A103,* 1988, pp. 167–184.
165. R. H. Jones and M. A. Friesel, "Using Acoustic Emission to Monitor Stress Corrosion Cracking," *Journal of Metals,* December, 1990, pp. 12–15.
166. W. W. Gerberich, R. H. Jones, M. A. Friesel, and A. Nozue, "Acoustic Emission Monitoring of Stress Corrosion Cracking," *Materials Science and Engineering, A103,* 1988, pp. 185–191.
167. Francis L. LaQue, *Marine Corrosion,* Wiley, New York, 1975, pp. 45–94.
168. Francis L. LaQue, "Corrosion Testing," *Corrosion* (L. L. Schreir, Ed.), Vol. 2, Wiley, New York, 1963, pp. 20.3–20.112.
169. "Standard Method for Conducting Corrosion Coupon Tests in Plant Equipment," *ASTM G4, Annual Book of ASTM Standards,* American Society for Testing and Materials, Philadelphia, PA, 1993.
170. R. A. Collacott, *Structural Integrity Monitoring,* Chapman and Hall, London, 1985, p. 331.
171. R. Viswanathan and S. M. Gehl, Ref. 127, p. 41.
172. J. F. Knott, Ref. 45, pp. 234–266.
173. T. S. Sudarshan and M. R. Louthen, Jr., "Gaseous Environment Effects on Fatigue Behavior of Metals," *International Materials Reviews,* Vol. 32, No. 3, 1987, pp. 121–151.
174. N. E. Frost, K. J. Marsh, and L. P. Pook, Ref. 101, p. 204.
175. Royce G. Forman and Tianlai Hu, "Application of Fracture Mechanics on the Space Shuttle," *Damage Tolerance of Metallic Structures, ASTM STP-842* (James B. Chang and James L. Rudd, Eds.), American Society for Testing and Materials, Philadelphia, PA, 1984, p. 112.
176. M. Vormwald, P. Heuler, and T. Seeger, "A Fracture Mechanics Based Model for Cumulative Damage Assessment as Part of Fatigue Life Prediction," *Advances in Fatigue Lifetime Prediction Techniques,* ASTM STP-1122 (M. R. Mitchell and R. W. Landgraf, Eds.), American Society for Testing and Materials, Philadelphia, PA, 1992, pp. 28–43.
177. Peter C. Riccardella, Arthur F. Deardorf, and Timothy J. Griesbach, "Fatigue Lifetime Monitoring in Power Plants," Ref. 176, pp. 460–473.
178. American Society of Mechanical Engineers, *ASME Boiler and Pressure Vessel Code,* Section III—Nuclear Power Plant Components; Section VIII—Pressure Vessels; and American National Standards Institute, ANSI B31.1—Power Piping.
179. J. C. Newman, Jr., E. P. Phillips, M. H. Swain, and R. A. Everett, Jr., "Fatigue Mechanics: An Assessment of a Unified Approach to Life Prediction," Ref. 176, pp. 5–27.
180. M. Vormwald, P. Heuler, and T. Seeger, Ref. 176.
181. Robert M. Engle, Jr., "Damage Accumulation Techniques in Damage Tolerance Analysis," Ref. 175, pp. 25–35.
182. M. G. Silk, "Defect Detection and Sizing in Metals Using Ultrasound," *International Metals Reviews,* Vol. 27, No. 1, 1982, pp. 28–50.
183. Steven Ashley, "Fatigue Alarm," *Mechanical Engineering,* January, 1992, p. 118.
184. Otto Buck, "Materials Characterization and Flaw Detection by Acoustic NDE," *Journal of Metals,* October, 1992, pp. 17–23.
185. Peter K. Liaw, "The Nondestructive Evaluation of Materials Properties," Ref. 183, pp. 10–11.

186. Robert E. Green, Jr., "Practical Applications of Nondestructive Materials Characterization," Ref. 184, pp. 12–16.
187. Robert E. Green, Jr., "Effect of Metallic Microstructure on Ultrasonic Attenuation," *Nondestructive Evaluation: Microstructural Characterization and Reliability Strategies* (Otto Buck and Stanley M. Wolf, Eds.), The Metallurgical Society of AIME, Warrendale, PA, 1981, pp. 115–132.
188. J. C. Shyne, N. Grayeli, and G. S. Kino, "Acoustic Properties as Microstructure-Dependent Materials Properties," Ref. 187, pp. 133–146.
189. Willie Hammer, Ref. 9, p. 264.
190. *Report of the President's Commission on the Three Mile Island Accident,* Pergamon, New York, 1979.
191. *Engineering News Record,* April 5, 1979, pp. 10–15.
192. Trevor Kletz, Ref. 126, pp. 93–101.
193. David Raj and Gary W. Roman, "Testing Power Generation Equipment," *Mechanical Engineering,* December, 1985.
194. Bill Evans, "Rugged Sensors," *Mechanical Engineering,* November, 1989, pp. 45–49.
195. Lee O'Connor "MEMS: Microelectromechanical Systems," *Mechanical Engineering,* February, 1992, pp. 40–47.
196. H. R. Jhansale and Donald R. McCann, "Fatigue Analysis Techniques for Vintage Steam Turbine/Generator Components," Ref. 176, pp. 474–489.
197. A. D. Kazanowski, "A Standardized Approach to Cost-Effectiveness Evaluations," *Cost-Effectiveness, The Economic Evaluation of Engineered Systems* (J. Morley English, Ed.), Wiley, New York, 1968, pp. 113–150.

Chapter

4

Failure Avoidance in Day-to-Day Practice

Product and Process Design

Priority and value readjustment

Previous discussions centered on operating equipment and systems and their continued safe and satisfactory performance. There the focus was upon strategies for avoiding failures in working components and extending usable lifetimes of existing systems. The approach in those discussions started with recognizing and identifying principal failure modes and fracture mechanisms, and how they came about. An awareness of susceptibilities of materials and characteristics affecting the likelihood and extent of attack by these failure mechanisms prompted inquiries of how these problems might be detected and monitored in existing structures, machines, and equipment. Damage mechanisms responsible for most failures in mechanical systems and the insidious failure modes that cause cracking in ordinary environments and under normal operating loads were emphasized. This did not imply that other failure mechanisms were insignificant or unimportant, only that the degree of severity and prevalence of these kinds of failures dictated that they should be given primary attention.

This chapter represents a shift in viewpoint, although the overall objective of failure avoidance remains unchanged. Attention in this section will be upon design and related concepts and approaches for minimizing product and process failures in everyday activity. In Chap. 3, the subjects of discussion—existing structures and operating systems and equipment—were "givens." The hardware had been designed, fabricated and constructed, and placed into operation beforehand. Our concern was in keeping it running safely, effectively, and efficiently, without failure and unscheduled outages, and in maintaining effective performance for the full extent of its useful life. Here, we start with a clean slate. The opportunity for exercising and applying failure avoidance strategies is much broader, as is the responsibility to ensure that they are exercised and applied. The range of possible problems, failure modes, and materials susceptibilities also is greater.

Chapter 1 described some of the changes in society's values over the past few decades, and reasons were discussed. These changes are reflected in many ways in our daily engineering activities and have had a major impact upon every industrial operation and organization. The effects have reached every engineer and others engaged in design, development, production and manufacture, marketing, sales, and distribution of products and services. One manifestation of this is evident in the product safety regulations affecting virtually every product or process and manufacturing operation. These are more than abstract documents to occupy the files of the firm's safety coordinator. They constitute new design standards, and it is the task of responsible engineers and managers to be familiar with them and to adhere to their provisions.

Working with standards has been a way of life for engineers from the beginning, as standards are indispensable to every measurement, instrument reading, and drawing. Design and production standards are the backbone of manufacturing and assembly operations, and standards govern the marketing and sale of every product.[1] In recent years, however, the number of standards involved in engineering activities has grown exponentially.

The Department of Defense Index of Specifications and Standards (DODISS) lists nearly 50,000 MIL standards and specifications. The 1993 Annual Book of ASTM Standards published by the American Society for Testing and Materials (ASTM) consists of 69 reference volumes containing 9,000 standards. The Standards Information Service, maintained by the Engineering and Product Standards Division of the National Institute for Standards and Technology, has a computer-indexed reference collection of over 20,000 standards, specifications, test methods, codes, and recommended practices prepared by nearly 400 U.S. technical, professional, and trade organizations.

As overwhelming as this proliferation of standards can be, the standards that can be the source of most apprehension for engineers, managers, and industrial executives are those that are formulated by our legal system—the courts of civil law. These may not appear on ASTM, American National Standards Institute (ANSI), American Society of Mechanical Engineers (ASME), or other engineering organization's lists, and many practicing engineers may not be aware of them. Yet, they just as surely specify characteristics and requirements of design, composition, construction, safety, reliability, risk–benefit aspects, communications between manufacturer and user, and the manufacturer's obligations to users and its customers.

Unlike industry-developed consensus standards, these generally do not suggest practices and procedures for achieving the desired end result. It is up to the manufacturer to not only work this out, but to be informed of their requirements, and to know that the standards exist in the first place. Court-produced standards certainly are not consensus standards but products of adversarial confrontations. One of their most onerous features is that they are retroactive and can include products made even before the standards were conceived. It may be blatantly unfair but binding nevertheless. And their interpretation in settlement of disputes is not for engineers to decide, but for attorneys and courts. As noted earlier in an excerpt of a leading court decision presented in Chap. 1 (Ref. 12 therein), standards established by law supersede industry standards.

The situation is similar to legislated regulations, as these are generally interpreted by courts and have the force of law. However, their origins are different from legal standards that stem from court decisions. They are products of elected representatives; in a sense, legislated regulations may be regarded as consensus enactments by society. In the recent furor over the nation's products-liability laws, it has been pointed out that even if all of these common-law-derived rules were abolished, the plight of the manufacturer would probably remain unchanged. This is said to be due to the fact that many of the same provisions have now been enacted into regulations having even broader application and scope.

Legislated regulations cover an extended range of manufacturing activities. Major ones include the Uniform Commercial Code; Magnuson–Moss Warranty Act; Consumer Product Safety Act (the Consumer Product Safety Commission also has administrative authority for federal hazardous substances, poison prevention packaging, flammable fabrics, and refrigerator safety legislation); National Electronic Injury Surveillance System (NEISS); Radiation Control for Health and Safety Act; Food, Drug and Cosmetic Act; National Traffic and Motor Vehicle Safety Act; Occupational Safety and Health Act; Clean Air Act; Toxic Substances Control Act; Noise Control Act; Water Pollution Control Act; the Solid Waste Disposal Act; The Energy Planning and Community Right-To-Know Act (Superfund Amendments and Reauthorization Act of October 1986—SARA Title III); and the list goes on.

Various departments and agencies of the federal government also prepare and issue their own regulations and standards or are responsible for administering those prepared by others. Some of these groups are the National Institute for Standards and Technology, Federal

Trade Commission, Federal Aviation Administration, National Institute for Occupational Safety and Health (NIOSH), Federal Highway Administration, Federal Railroad Administration, Environmental Protection Agency, The Department of Defense (Military Specifications and Standards), National Transportation Safety Board, and the Nuclear Regulatory Commission.

The thrust behind this formidable mountain of regulations and the rationale for imposing the burden upon industry are clear from the following excerpt of a report by the National Commission on Product Safety created by Congress under Public Law 90-146, and which subsequently established the U.S. Consumer Product Safety Commission (CPSC):

> The law has tended in recent years to place full responsibility for injuries attributable to defective products upon the manufacturer....But beyond his liability for damages, a producer owes society-at-large the duty to assure that unnecessary risks of injury are eliminated. He is in the best position to know what are the safest methods, materials, construction methods, and modes of use. Before anyone else, he must explore the boundaries of potential danger from the use of his product. He must be in a position to advise the buyer competently how to use and how to maintain and repair the product.[2]

Although the scope of the CPSC is consumer products, the above statement describes the philosophy behind much of the safety legislation enacted during the last few decades. As a point of interest, note the clear parallel between the above statement and the excerpt of a court decision cited in Chap. 1 (Ref. 11 therein). The similarity is more than coincidence.

As suggested before, what we are seeing in all of this is the increasing intolerance of society for defective products and, more than that, its backlash to damages and injuries traceable to manufactured products and modern technology. The far-reaching and compelling influence upon industry is apparent from the regulatory climate facing engineers and managers today. There are few, if any, remaining loopholes for escaping recriminations over product failures and virtually all other failures, whether they involve the user of a farming implement, a factory worker in a manufacturing or process plant, or victims of industrial disasters, vehicular defects, or hazardous waste spills.

This is, perhaps, an opportune time to recall our broadened definition of "failure" of Chap. 1 and to note that it covered a great deal more than physical fragments and wreckage. Cracks and material fractures and their various damage mechanisms do play a major role in engineering failures and are principal factors in prolonging useful lives of operating systems and in avoiding equipment failures. But in our engineering activities we must strive to avoid *all* of the elements and facets of our original definition, as seeds of litigation lie within many so-called minor incidents.

During recent years, costs of engineering failures have escalated rapidly, as society's intolerance for them has been felt in many ways. It is felt not only through the direct cost of defending lawsuits and the expense of product recalls and plant shutdowns, but also from lost profits from disenchanted customers, increased insurance premiums, eroded investor confidence, negative public image, and intensified bureaucratic interference.

What does all of this mean in terms of engineering design? Traditional design techniques—the engineering concepts, mathematical formulas, mechanical analyses, and physical relationships—are little affected, although their framework of application will change as new disciplines such as fracture mechanics and ergonomics are added. What must change appreciably, however, are designers' priorities, attitudes, and their responsiveness to the increasingly critical need for minimizing failures, which has prompted the flood of new legal standards and legislated regulations and costly litigation and product recalls.

This means that engineers must become knowledgeable and proficient in new subjects such as failure analysis, systems techniques, hazards analysis, failure mode and effects analysis (FMEA) and fault-tree analysis (FTA), design reviews, damage-tolerant design, fail-safe and safe-life concepts, risk–utility analysis, and remedial practices, with greater emphasis upon codes and standards, instruction manuals, warning labels, and warranty issues. There must be more interdepartmental communication and coordination to ensure an integrated approach to product safety and reliability. These issues are acute in companies having product lines that are vulnerable to damaging failures and their consequences. However, no manufacturing firm today can afford to remain complacent in the light of these new developments, no matter how immune it appears to have been so far.

All of this is not easily accomplished, as most of these developments and requirements evolved outside of industry's traditional standards-making organizations. Nevertheless, engineers must supplement their technical information monitoring system with information on the implications of new court decisions and regulations. Sources for this information may lie outside the usual engineering disciplines and familiar journals, but are accessible. There is an increasing number of information services devoted to monitoring these developments. They provide subscribers with periodic digests on a biweekly or monthly basis, and some are available on computerized databases that can be accessed at any time from an engineer's office. The company library may already be a subscriber to some of these information sources.[3–11]

Unfamiliarity with or inconvenient access to information sources on legal developments and regulations affecting engineers' decisions and activities is, of course, no excuse for not being informed. In the aftermath of a damaging incident affecting the company, evidence of ignorance in these matters could be interpreted by courts as negligence in and of itself, as disregard for the safety and welfare of others, and repudiation of professional responsibility.

Responsible individuals will see to it that they become aware of these matters, whether they do so through seminars, courses, or reviews of appropriate literature. Personal involvement in a lawsuit stemming from a failure of some kind certainly is an educational experience, but this is not the way engineers should obtain their information about the legal implications of what they do, do not do, or should have done. At this fairly well-advanced stage of evolution of civil law affecting manufactured products and related issues, basic concepts are well established and should be understood by every engineer. They should be as familiar to the mechanical engineer as Newton's second law; it is not a task that is delegable to others.

New design criteria

Until only a few decades ago, injured victims and those who had suffered damage from engineering failures had to prove negligence on the part of the manufacturer to sustain legal action. Negligence relates to conduct, and in a negligence action over a product failure, the *manufacturer* is on trial. To prove negligence, an injured plaintiff must show that the defendant manufacturer owed him or her a duty of care and breached that duty, and this fault must be relevant in causing the alleged harm. *Fault,* in a legal context, is simply behavior that fails to measure up to the standard imposed by law.

Negligence liability can be incurred in what is done or what is said, and the legal elements and procedures required for plaintiffs to succeed in such cases can be complex, although this basis for civil action is still very common. An important element that distinguishes negligence actions is that the burden of proof (of duty of care, its breach, and causation) lies squarely upon the injured plaintiff.

Since the 1940s, however, the development and widespread adoption of the legal concept of *strict liability* in most jurisdictions of this country has relieved the plaintiff of the burden to prove the elements necessary to sustain a negligence action. In strict liability, the plaintiff need only prove that the product that caused his or her injuries or damage was defective and was in this condition when the manufacturer placed it into the stream of commerce. This concept represents a shift in emphasis from the manufacturer's conduct (negligence) to the product's condition (strict liability). This distinction is clearly noted in the following excerpt of a 1974 court decision:

> In a strict liability case we are talking about the condition...of an article...while in negligence we are talking about the reasonableness of the manufacturer's actions....[12]

Rationale for the shift in emphasis is explained in this excerpt from a 1978 California Supreme Court decision:

> [O]ne of the principal purposes behind the strict products liability doctrine is to relieve an injured plaintiff of many of the onerous evidentiary burdens inherent in a negligence cause of action....
>
> [T]echnological revolution has created a society that contains dangers to the individual never before contemplated. The individual must face the threat of life and limb not only from the car on the street or highway but from a massive array of hazardous mechanisms and products. The radical change from a comparatively safe, largely agriculture society to this industrial unsafe one has been reflected in the decisions that formerly tied liability to the fault of the tortfeasor but now are more concerned with the safety of the individual who suffers the loss....
>
> [T]he change in the substantive law as regards the liability of makers of products and other sellers in the marketing chain has been from fault to defect. The plaintiff is no longer required to impugn the maker, but he is required to impugn the product.[13]

In a strict liability action, the *product,* essentially, is on trial, as its defectiveness is the key element. It is not *liability without fault*; it is liability without the need for proving fault. Much of this distinction is legal formalism, however, since it is the manufacturer who still is responsible for the defective condition of the product and it is the manufacturer who must pay damages in an adverse decision. Strict liability offers evidential and procedural advantages to the injured party and, therefore, an easier and more direct route to recovery of damages and financial compensation—the goal of products-liability actions. And it imposes new design criteria upon the product manufacturer.

Targets of these lawsuits can be anyone in the product design, manufacture, and distribution chain. The range of possibilities is evident from the following definition of *manufacturer* from a Senate proposal for uniform products-liability legislation:

> [M]anufacturer means (A) any person who is engaged in a business to design or formulate and to produce, create, make, or construct any product (or component part of a product), including a product seller, distributor, or retailer of products with respect to any product to the extent that such a product seller, distributor or retailer designs or formulates and produces, creates, makes or constructs the product before that product seller, dis-

> tributor or retailer sells the product; or (B) any product seller not described in clause (A) which holds itself out as manufacturer to the user of the product.[14]

Although this large cast of potential defendants offers plaintiffs many options, the choice often boils down to those having sufficient financial resources—the "deep pockets"—to pay damage claims, as economic compensation is the name of the game. This does not mean that they are the only ones named as defendants, however. The trend is to include all who may have been associated with the failed product. There are procedural advantages to plaintiff attorneys in this as it increases the availability and accessibility to information that they need in pursuing the case.

A plaintiff's success in sustaining legal action in strict liability hinges upon proving defectiveness of the product alleged to have been the cause of the damage or injury. Therefore, this issue requires some consideration here and, understandably, occupies a great deal of attention during litigation. A reliable definition of *defect* has eluded courts and attorneys for years. This is probably due to the constantly evolving nature of the concept.

A definition suggested by a products litigation attorney illustrates the difficulty of defining this concept and the sentiment of plaintiff attorneys: "A defect is anything that is wrong with a product—which is causally connected with the injury complained of."[15] Defectiveness in a product or article may result from faulty design, faulty manufacture, or faulty communication. Here we are dealing with much more than intergranular cracks and missing cotter pins. For example, a product may be branded *defective* by a court because it did not bear a proper warning label.

User expectation can be an important element in proving product defectiveness. A user may understand that use of a given manufactured product incurs a certain degree of danger. This ordinary awareness prompts attention to be paid to these characteristics. For example, an injury that might result from careless or thoughtless use of an ax probably would not be a basis for strict liability action as the primary purpose of an ax is to cut wood. It must be sharp to do so; consequently, it has an inherent but necessary propensity for harm. But suppose that, in its usual use in cutting wood and as a result of a metallurgical flaw, a sharp-edged metal fragment breaks off the ax and strikes the user. Then the harm associated with use of that ax is greater than the user might expect and would likely be a basis for sustainable action in strict liability. This is known as the *user expectation test* in determining product defectiveness. Not all situations are as uncomplicated as this simple example, however.

Alleged *design* defectiveness in a strict liability action, or *design* negligence in a negligence action, is not straightforward. There is little distinction between the two because design issues revert to questions of conduct, a negligence characteristic. The plaintiff is at a disadvantage in proving design defectiveness because of lack of information on technical alternatives, safety options, and costs. In such cases, courts have applied rules governing design defectiveness that are based upon user expectation or some form of cost–benefit analysis, as explained in the California case excerpt noted below:

> [A] product may be found defective in design so as to subject a manufacturer to strict liability for resulting injuries, using either of two alternative tests. First, a product may be found defective in design if the plaintiff establishes that the product failed to perform as safely as an ordinary consumer would expect when used in an intended or reasonably foreseeable manner. Second, a product may alternatively be found defective in design if the plaintiff demonstrates that the product's design proximately caused his injury and the defendant fails to establish, in light of relevant factors, that, on balance, the benefits of the challenged design outweigh the risks of danger inherent in such design.[16]

Standards that courts apply in determining design defectiveness are illustrated in a 1983 New York case where seven elements in a risk–benefit analysis are described:

> The question for the jury, then, is whether after weighing the evidence and balancing the product's risks against its utility and cost, it can be concluded that the product as designed is not reasonably safe....In balancing the risks inherent in the product, as designed, against its utility and cost, the jury may consider several factors....Those factors may include the following: (1) the utility of the product to the public as a whole and to the individual user; (2) the nature of the product—that is, the likelihood that it will cause injury; (3) the availability of a safe design; (4) the potential for designing and manufacturing the product so that it is safer but remains functional and reasonably priced; (5) the ability of the plaintiff to have avoided injury by careful use of the product; (6) the degree of awareness of the potential danger of the product which reasonably can be attributed to the plaintiff; and (7) the manufacturer's ability to spread any cost related to improving the safety of the design.[17]

These excerpts were chosen from court decisions in jurisdictions of high visibility that have played a major role in the development of strict liability principles. They are useful in offering insights into standards that courts apply to engineered products. Design standards used by engineers should be no less demanding.

Manufacturing defects generally do not present the problem of proof encountered in alleging design defectiveness, as it is usually not difficult to show that the defective product was not what the manufacturer intended it to be. When the manufacturer realizes and admits this, the dispute is usually over. This often leads to an early negotiated settlement, the method of disposition of most products cases.

As noted briefly before, defective *communication* associated with a manufactured product can also be a basis for liability. These cases usually involve failure to warn of some danger associated with use of a product. In the ax example described earlier, users would be expected to know that axes are sharp and therefore dangerous to use. But all users might not be aware that use of the back surface of an ax as a hammer to strike hardened steel implements can cause fragments to fly off. This might not be caused by defects in the material or manufacture of the ax but a consequence of work hardening and other properties of steel tools. This situation cannot be effectively guarded against or prevented; therefore, it should be warned against and usually is.

Defectiveness of a product or object that has propensity for harm usually is determined by whether it is *unreasonably* dangerous. This characteristic can be the result of a design defect that is not evident to normal users or from lack of some safety feature and is evaluated by assessing the degree of preventable danger. Unreasonable danger in a product is also assessed by evaluating its utility and whether its usefulness is outweighed by the probability of harm that might result from its use or foreseeable misuse or abuse.[18]

Some essential products and devices would be considered *unavoidably unsafe,* as are some vaccines, drugs, and industrial chemicals. However, they would probably not be considered unreasonably dangerous, and therefore the target of a strict liability claim, unless they were packaged inadequately or failed to have appropriate and prominent warnings of their risks, or did not comply with some regulatory requirement on restricted distribution (drugs dispensed without a physician's prescription, for example).[19]

The legally mandated requirements described, and others that are under consideration, constitute design criteria that must be applied to all applicable engineered products and systems. Courts use the generic term "product" to describe manufactured objects responsible for alleged damage or harm. But its scope can be extremely broad, as noted in the following definition:

> "[P]roduct" means any object, substance, mixture or raw material in a gaseous, liquid or solid state which is capable of delivery itself, or as an assembled whole in a mixed or combined state or as a component part or ingredient, which is produced for introduction into trade or commerce, which has intrinsic economic value, and which is intended for sale or lease to persons for commercial or personal use.[20]

Design requirements mandated by law and regulations are most effectively implemented on a corporate-wide basis through a top management directive, as it is difficult for individual engineers to swim against the tide if corporate management tends to ignore such issues. Any program of failure avoidance must include them, however, and the best starting point is an overall appraisal of the firm's liability potential in the light of recent court decisions and regulatory developments. These issues should have been included in the appraisal of the need for implementing failure avoidance strategies, discussed in an earlier chapter. If the "rules" affecting how business is conducted are changed, it is essential that policy adjustments, operational modifications, and even product designs and formulations are also changed to accommodate them. Failures of businesses are frequently traceable to unwillingness to adapt to new standards. Even worse are failures due to ignorance of the standards or incorrect assessment of their relevance and probable impact.

It is necessary for a company to frequently reassess potential effects of recently enacted regulations or court decisions to not only keep abreast of these developments but to ensure that its products and conduct are consistent with these requirements. These reassessments should include reviews of product policies and product-line vulnerability to failures and products lawsuits. Evaluations in the light of these requirements should be made of advertising claims, instruction manuals, packaging, warning labels, and sales policies. They should also include surveys of materials and constituents used and their possible hazardous characteristics, with attention to stored materials, process effluents, by-products and wastes, and disposal practices. Examples of once-commonly-used materials and reagents that are now considered hazardous include carbon tetrachloride, benzene, asbestos, formaldehyde, and polychlorinated biphenyls (PCBs). There are many others.

Products and practices that at one time were profitable and solidly entrenched in the corporate product line may require modification to avoid future liability problems or elimination altogether. Materials considered inert under ordinary conditions may become unstable in time and hazardous, or release toxic fumes when decomposed or burned, and become targets of litigation. An entire product line may be considered defective if its risk of injury outweighs its utility or social benefits. The toy industry is particularly vulnerable to these kinds of problems, but other product types can be similarly affected (e.g., household products, insecticides, paint, and fabrics). These considerations can have substantial impact upon a firm's manufacturing activity, its product line, and its design approaches, among other business and operational aspects.

Although compliance with mandatory regulations, voluntary standards, and industry practices does not confer immunity to lawsuits or guarantee trouble-free product life, in court, it does demonstrate evidence of due care. However, lack of compliance can be presumptive evidence of negligence. Code compliance is regarded by courts as a floor, not a ceiling, and offers limited defense during litigation. The weight accorded by law to code compliance is typified by the following statement by the National Commission on Product Safety:

> Since approval of a voluntary standard requires a consensus of those concerned, the participating manufacturers tend to agree on safety levels that are least costly and troublesome....If a manufacturer fails to meet standards his own industry considers minimal, the evidence should be conclusive proof of fault. If a product does meet the industry standard, this evidence of due care may be rebutted by proof that the entire industry has been lax. Safety standards of industry...seldom satisfy objective requirements....[21]

Responsibility of today's designers obviously extends well beyond technical issues. With broadened definitions for *failure* and *defectiveness,* designers must reappraise and readjust their standards and criteria to accommodate them to avoid potentially serious problems.[22–27] This must be done at each stage of design. Greater emphasis must be given to users' physical and mental capacities, to users' needs for the product and their motivations for obtaining it, to their expectations for what it can do, and to the environment.

It will be apparent that much of this is beyond the control and responsibility of the designer. For example, advertising may suggest uses or levels of performance that are potentially hazardous. Production personnel may make material substitutions or use alternative joining processes or assembly methods. Sales representatives may have their own ideas and opinions of product applications or performance capability that will offer a competitive edge over products or systems marketed by others. Such well-meaning but potentially detrimental activities and attitudes can set the stage for trouble down the road if allowed to persist or be implemented. This is one of the reasons for design reviews.

Design of a product, component, structure, manufacturing or chemical process, or an entire plant or engineering system is a critically important activity for a number of reasons. During this stage, commitments are made and configurations established that fix its form, substance, and destiny. The design stage identifies the market and user; determines the product's function, principles of operation, and features; specifies materials and fabrication practices; sets performance standards and capabilities; and establishes inspection criteria and methods. It also determines much of its potential for harm.

Therefore, the design process must combine hazards assessment and control procedures with traditional engineering tasks. Admittedly, not all products, components, and systems require extensive hazards elimination and control procedures, as some have minimal hazard potential. But this determination should not be made intuitively at the outset with the result that product safety issues are dismissed or formal hazards assessment exercises are bypassed. All new and modified designs should be subjected to a formalized review process for these determinations, whether they appear to have a hazardous potential or not.

The following sections summarize strategies that have been devised and developed to accommodate the new design criteria imposed through court decisions and legislated regulations. There is no particular significance to the order in which they are presented. Each design project will have its own specific requirements and characteristics that will differ from all others, requiring its own unique treatment. The practices and strategies described are but a few of those that have been and are being applied. Because of space limitations, coverage must be limited in number as well as in depth. Our goal is not a detailed design practice manual but a generalized description of guidelines and approaches that have been successfully applied in dealing with these new and difficult issues.

Assessing hazards in design

Four principal tasks characterize most formalized design safety practices: (a) hazards identification, (b) hazards elimination, (c) hazards control, and (d) disposition of persisting hazards. It will be noted that there are similarities in terminology and assessment procedures among techniques used during design of new components and those used for evaluating existing operating equipment for failure susceptibility, discussed in Chap. 3. This should not be surprising as these are analytical tools having general utility for a wide range of applications.

Hazards identification. The law imposes an obligation upon manufacturers to see to it that users and reasonably foreseeable bystanders, surrounding property, and the environment are not harmed or damaged by defective products or systems. It is a perpetual obligation as it extends throughout their existence and even after their useful lives are over. Engineers' concepts of defectiveness, as briefly discussed earlier in this chapter, must be broadened to encompass safety considerations and defectiveness criteria used by courts to evaluate engineered products.

In addition to faulty design, defectiveness can include improper choice of materials and constituents, lack of safety devices, inadequate warnings and instructions, faulty manufacture and assembly, and misleading advertising and sales practices. And it can include failure to deliver expected benefits and performance.

Defects may arise during any stage of production, fabrication or construction, installation, maintenance, and disposal, where safety is compromised. This can include design, manufacturing, assembly, materials selection and procurement, use of component parts, inspection, packaging, instructions, warnings, promotion, warranties, certification, and testing.[28]

During hazards identification evaluations, users and those likely to come into contact with the product and who may be injured by it should be identified, along with circumstances of use, contact, and environment. These people may be handicapped, elderly, children, and illiterate. All of them must be contemplated in identifying hazards.

A variety of hazards identification techniques have been used during these evaluations.[29] Their choice depends upon the product's characteristics and many factors, and no single method is best. They are complementary and it is desirable to use several techniques. Extensive lists of hazardous conditions, characteristics, and processes have been prepared and should be consulted.[30–32] However, primary hazards associated with a given product, process, or system can usually be readily identified, although some may not be obvious.

Particular attention should be given to substances that may be harmful or are toxic and those having explosive potential (reactive, pyrophoric, carcinogenic, flammable, corrosive), operations involving high pressures and temperature extremes (both high and low), initial startups of experimental prototypes or modified systems, and "shakedown" runs of new equipment and systems.

Hazards analysis is usually the first step that defines hazards and identifies hazardous elements (see hazards analysis discussions in Chap. 3). Various techniques may be used during this and other stages, such as "what if," criticality, and HAZOPS (hazards and operability study) analyses.

Using product checklists and operational or process flowcharts and diagrams, "what if" scenario analysis considers events or conditions that may deviate from normal or anticipated practice or response.[33] They can include system faults, component failures, environmental effects, personnel behavior, and any number of conceivable situations. The objective in raising these questions is to challenge each system component and each operational aspect and force analysts to address the situation and formulate a reasonable reply. In this way, hidden hazards are often brought to light.

Criticality analysis evaluates the damage potential of faulty or defective components and subassemblies as a function of their probability of failure and its impact upon the integrity of the system or successful attainment of the desired goals.[34] It is an ordering or prioritizing technique that provides guidance in determining items that require primary attention. Often, the procedures are incorporated into FMEA studies and are designated failure modes, effects and criticality analysis (FMECA) (see Ref. 31 of Chap. 3). Its results are useful, not only in identifying components requiring more intensive study for failure avoidance purposes, but also in identifying those items that should be closely monitored or specially protected during manufacture or throughout their useful life and perhaps beyond. This information is also beneficial in the cost-effective allocation of resources in determining which items require most attention and, therefore, the greatest expenditures, in minimizing failures.

HAZOPS is an extended and more rigorous version of FMEA that includes operational factors as well as equipment failure modes.[35] The technique focuses upon possible operational deviations from design conditions, using standardized guide words to evaluate system consequences of those deviations. It requires detailed knowledge of the system involved and all its operational variables and can be time consuming to conduct properly. The method analyzes operating aspects of concern through applying the guide words as a means of deriving process deviations for every system parameter, such as stress level, temperature, pressure, flow rate, etc. It then evaluates consequences and causes of these deviations. HAZOPS uses a spreadsheet format and its output is in the form of corrective recommendations. Other techniques described in Chap. 3, such as FMEA and FTA, are often useful in defining hazards and their causative or triggering elements, in addition to the hazards analysis methods described above.

Checklists can be indispensable in facilitating hazards identification. These may be modeled after formats used by others for similar products and applications, or prepared specifically for the intended use. As a minimum, they should address the following issues:

1. Purpose of the product, its intended use, and foreseeable misuse and abuse
2. Product elements and possible associated hazards and their severity
3. Product–user interface during normal use and foreseeable misuse and during maintenance and repair
4. Survey of toxic substances in light of regulations from viewpoints of manufacture, sales, storage, and disposal
5. Adequacy and comprehensibility of user guidance and instruction

6. Warning labeling requirements and use
7. Required agency certifications or approvals (e.g., Underwriter's Laboratories or the Food and Drug Administration)

Other issues prompted by industry convention, custom, or standard practices should also be included.

Hazards elimination. Identified hazards should be eliminated and "designed out" of the product or system, if possible. This is, by far, the best and preferred option. The old adage says it well: "Quality, reliability, and safety should be designed in, not inspected in." However, it usually comes down to a matter of cost, and this is where manufacturers can get into trouble. It is therefore important to consider carefully criteria used by the courts in design cases in assessing product defectiveness.

The two principal tests described in a previous section (Refs. 16 and 17) were the user expectation test and a risk–benefit or risk–utility test. Courts can be hypersensitive to product features they consider defective (e.g., inadequate user protection or faulty assembly or material) and which the manufacturer judged too costly to change. Such situations can prompt courts to impose exorbitant punitive damage awards. Costs associated with eliminating hazards are obviously important to the manufacturer; however, there should be no evidence that economic considerations dictated design decisions and policies affecting product safety or were dominant.

Risk–benefit analyses, like other analytical methods, are best conducted using standardized formats. This ensures that all elements are included, provides a tangible basis for containing and documenting input and results, and serves as a record of the activity. Guidelines may be obtained from Ref. 17 or other sources.[36]

Basic questions to consider would include objective appraisals of the product's usefulness. Product alternatives should be identified along with their technical feasibility, costs, and other effects. Other product aspects that should be evaluated would be risks, their severity, and related user expectations; the feasibility of eliminating risks; risk minimization approaches and their effectiveness; costs for risk elimination; and impact of these factors upon marketability of the product. Weightings may be assigned to the various elements and the product rated along with reference products to help assess the results and to establish guidelines for making decisions.

As a minimum, all designs should conform to standard industry safety practices and relevant state-of-the-art as well as internal corporate standards. As already mentioned, their compliance does not impart immunity to liability claims, but failure to meet such standards can be damaging.

Hazards control. This step would be considered only after the preceding step was completed and it had been determined that although potential hazards persisted, usefulness of the product was such that the associated risks were acceptable. Nevertheless, every reasonable precaution should be evaluated for controlling the risk. Guards, interlocks, protective enclosures, and other such devices and approaches might be used. Although manufacturers may be tempted to offer these devices as options or accessories, this approach can be an invitation to liability, depending upon the hazard, the product, and jurisdiction involved. A preferred approach is to make safety features integral parts of the design and standard for the equipment and system.

Warnings and instructions. If hazard potential persists despite all attempts to eliminate it or guard against it, and the product design has survived the scrutiny of a risk–utility analysis, the danger must be warned against. It has been demonstrated repeatedly that warnings do not compensate for unsafe designs. Therefore, any reliance placed upon this approach should be carefully appraised in consultation with legal counsel.

With increasing technological sophistication of products, the gap in level of understanding between product designers and users has widened considerably. What may be readily apparent and obvious to trained professionals can be obscure to most laypersons and product users. Users may believe that they fully understand the operating principles and associated risks; yet, most often, they do not and can misjudge risks sufficiently to be vulnerable to harm or injury. It is up to the manufacturer and designer to effectively bridge this gap to avoid damaging consequences.

There is an opportunity for conflict here, as the customer or user is frequently exposed to repetitious manufacturers' advertising claims regarding product performance and other features. These create strong impressions, as they are designed to do, to induce prospective buyers to purchase products. And they do more than this, they also create expectations based upon the advertising claims. Sometimes, these expectations are inconsistent with safe use of the product. As noted earlier, product *defectiveness* can stem from user expectations, and these can come from advertising claims, television demonstrations, and sales pitches.

When the prospective buyer obtains a product, he or she may have been preconditioned to expect performance and other results that are unrealistic or impossible. Consequently, manufacturers' instructions and warnings accompanying the product that tend to moderate or temper these expectations are often disregarded and ignored. This makes the task of devising effective instructions and warnings particularly difficult, and points up the importance of a coordinated effort to minimize liability problems.

Much has been written on product warnings and labeling requirements.[37–39] Most authorities feel it is advisable not to rely upon user manuals for hazard warnings, as many do not read these manuals, or do so only after trial runs to check out the product. Several basic guidelines suggested in the references for use of product warnings call for them to (a) be prominent, conspicuous (located near the hazard), and unambiguous, (b) be plainly descriptive of the hazard, (c) clearly state the severity of possible harm, (d) advise on how to avoid the hazard and the likely result of ignoring the warning, and (e) be permanent.

The warnings must be comprehensible and applicable to all foreseeable users of different ages, backgrounds, sizes, sexes, and capabilities. The warning label designer must place himself or herself within the shoes of the user to develop the proper placement and content consistent with the most probable use pattern.

Label size, its wording and type style, color, pictorial symbols, and layout all are important design considerations. Many regulations, particularly those dealing with toxic materials and hazardous substances, spell out labeling requirements and these should be followed as applicable. Warnings are an especially critical aspect of failure avoidance in product design because they are, essentially, a last resort and their very presence confirms the existence of an unavoidable hazard.

User instructions should carry the same safety theme throughout and be consistent in all respects with warnings affixed to the product and elsewhere. Adequate instructions can play an important role in avoiding misuse and abuse and should be clearly written and oriented toward the user's standpoint.

If the product is contemplated to require assembly by the user, the instructions should be unambiguous and clear, with adequate and readable pictorial diagrams to ensure correct assembly. Errors that can incur risks of product failure or user injury should be clearly pointed out. It would be preferable to avoid these risks through factory assembly of critical components.

Instructions should be comprehensible to all users, and this might require versions in several predominant languages. Also, all instruction manuals should be test-read by laypersons unfamiliar with the product and its technology to evaluate their comprehensibility and accuracy. All discrepancies should be corrected before final approval and release.

Design reviews

Because it is more cost-effective to identify and correct deficiencies while the product is still on the drawing board and before the manufacturing phase, the design review has become a widely accepted practice. Such reviews may be conducted at any time, but formalized design reviews most frequently are scheduled at three stages of the product's gestation: (a) the preliminary design review, (b) the intermediate, or prototype, design review, and (c) final design review. The objective is to ensure an optimal design, considering all product elements necessary for fulfilling the customer's needs, and that the product yields a satisfactory profit when priced competitively.

Design reviews are conducted to ensure that the product or system will be safe and reliable, that costs and materials are optimal, and that applicable regulations have been satisfied. The design review is an advisory function established to provide a forum to communicate and measure achievement of these objectives from the standpoints of performance, total cost, safety, reliability, producibility, environmental effects, maintainability, serviceability, human factors, customer needs and expectations, and pertinent legislation and litigation, including personal injury, property damage, and environmental damage.[40] It is a scheduled, systematic review and evaluation by qualified personnel not directly associated with the product design activity but who, collectively, are knowledgeable in and have responsibility for all elements of the product throughout its life. The design review is primarily a management tool, in contrast to the analytical tools described above.

Choice of participants depends upon the product and stage of the design review, but they should include knowledgeable representatives of product management, design, safety, procurement, production, marketing, legal, sales, and advertising departments. These may be supplemented, as necessary, with specialists having expertise in reliability engineering, maintainability, tooling, packaging and shipping, field service, quality control, human factors, environmental effects, patents, licensing, and those familiar with pertinent legislation and pending litigation. Since a large number of participants can create an unmanageable situation, smaller working groups assigned to subsystems may be desirable for large projects involving many disciplines. These can be integrated later into the total picture. Design review checklists are a practical necessity.[41] Product reviews are not conducted to criticize the design staff but to offer constructive comments on the design itself.

The preliminary design review is usually held during the conceptual and planning stage before the initial design is formulated. A primary purpose of this session is to establish interdepartmental communications. It also evaluates the concept and offers participants an opportunity for airing misgivings or apprehensions about it, or other comments, and does so at a time when changes and tradeoffs are readily made. Questions that are raised and unresolved issues may be identified for study and analysis using various techniques such as hazards analysis, FMEA, FTA, and other methods.

The intermediate design review is convened after preliminary layouts are completed but before detailed production drawings are made. It is conducted to evaluate the modified design against performance requirements. It is the time to review results of the analytical and tradeoff studies and outcomes of any prototype tests, to discuss mechanical and electrical designs, and to evaluate tooling needs, packaging requirements, and installation and maintenance manuals.

The final design review takes place when the design is ready for release to production, and after material lists and production drawings are prepared. All previously identified problems and adequacy of their resolution should be reviewed, and all test data, results, and reports on subsystems should also be available for review. Manufacturing feasibility can be an important issue and often is the subject of a separate review at this stage.

An essential product of design reviews is a final report. It should reflect all deliberations, decisions, and tradeoffs and their rationale and include analytical results, test outcomes, dispositions, and resolutions of questions and issues. All subsystem reports and input from consultants and other disciplines should be included.

The existence of a coherent and comprehensive document describing the product review process in detail can have significant value in demonstrating responsibility in striving to achieve a safe design. In addition to this benefit as a litigation defense tool, the design review process results in a superior product design with respect to safety, reliability, materials and manufacturing efficiency, maintenance, and cost. Through its use, the need for costly changes and modifications during production and other stages is drastically reduced.

Damage-tolerant design

This is one of several design concepts where there has been confusion over definitions. This is a relatively insignificant issue but one that can lead to misconceptions and misunderstandings by those not directly associated with these subjects on a daily basis. Other related terms are *safe-life* and *fail-safe,* to be described later. The brief summaries here do not attempt to resolve the semantic confusion, nor will they be completely definitive. The design concepts are sound, no matter how their descriptive terms are defined, and they have proven themselves useful for well over two decades in a variety of critical applications.

Cyclic loading. During recent decades (1960s–1970s), efforts to improve material properties for more efficient (higher strength to weight) designs of aerospace and other critical structural components were occasionally marred by failures. As noted in previous discussions, components fabricated of higher performance materials are more susceptible to catastrophic damage mechanisms. This is largely due to their higher operating stress levels; but adverse material microstructures and fracture properties also contribute.

During analysis of these failures, it became evident that sole reliance upon nondestructive examination (NDE) for detecting failure-initiating flaws was risky, at best. Through hindsight, it is now evident that critical flaw sizes in these high-strength materials probably were near or even below NDE detectability limits. This had been demonstrated by failures that occurred during initial trials and test runs of previously inspected components even before the equipment was placed into service.

The fatigue failures initiated at imperfections, flaws, and material discontinuities; some were inherent in the material and others were introduced during manufacture, assembly, or at some other stage. These exist in virtually every component, regardless of the care and time expended to eliminate them. It became apparent that an approach was needed to accommodate their presence. That is, acknowledge that their existence is inevitable, but yet design the structure or component in such a way that this damage would be tolerated during its projected or required lifetime.

This objective would have appeared overly ambitious, wildly optimistic, and impractical until the development of linear elastic fracture mechanics principles. This is an analytical tool that can predict structural behavior of flawed components. It does so through quantifying crack propagation properties of materials and effects of various flaw sizes and orientations. (A review of these subjects in Chap. 3 and related reference materials is recommended.)

The fracture mechanics approach to safe design of a structure or component containing acknowledged but undetected flaws and cracks requires several items of information:

1. Size of the largest crack that could escape NDE detection
2. Conditions that could cause this crack to grow (e.g., stress, environment, loading configuration)
3. Extent of crack growth under these conditions
4. Ultimate size of that crack for onset of catastrophic failure (i.e., unstable crack propagation) under the operating stress.[42]

As discussed earlier, the stress-intensity factor K controls crack growth. It defines the relationship between crack size and the level of stress required for it to grow. The limiting, or critical, value of K, designated K_c (or K_{Ic} for the crack-opening mode I, of usual interest) denotes the value of K at which crack growth becomes unstable. This limiting value is regarded as the material's fracture toughness, a material property, and may be found for common engineering materials in handbook tables.[43]

Controlling factors in a *damage-tolerant* design analysis are (a) the residual strength, or load-carrying capacity of the component in the presence of cracks, which establishes the fracture stress for a given flaw size, and (b) the number of loading cycles needed for a subcritical-sized flaw to grow to critical size (i.e., when $K = K_c$).

For these parameters to apply rigorously, behavior of the material in the vicinity of the crack tip is assumed to be elastic (no plastic deformation), although a plastic-zone correction factor Q is often applied to accommodate actual conditions more closely. Values for K for various crack geometries may be obtained from handbooks or derived from published equations.[44,45] With this information and fracture toughness values, it is possible to determine critical crack size for a given material and structural configuration.

The size of the largest crack that can escape NDE detection by the most sensitive method must either be determined experimentally or obtained from published values.[46–49] Assuming an initial subcritical flaw size, derived from demonstrated NDE capabilities, that is present in the component, the number of cycles required for the flaw to grow to critical size may be calculated from expressions that relate fatigue crack-growth rate to the stress-intensity-factor range for the given material.[50]

Published da/dN versus ΔK data can also provide rough approximations. ΔK is the range of stress-intensity-factor variations ($K_{max} - K_{min}$) during cyclic loading, and da/dN is the cyclic stress crack-growth rate expressed in mm/cycle. More reliable data may be obtained through conducting crack-growth studies (da/dN tests) for specific material–environment combinations.

ΔK controls the rate of fatigue crack growth and thereby cyclic life. Having determined the size of the largest initial crack that could be present in the component (after NDE inspection) and knowing the size that it could become (via K_{Ic}) before the onset of unstable fracture, the limits for determining the number of elapsed stress cycles needed for the change in crack length are fixed. Then, the appropriate crack-growth-rate expression can be integrated between these limits to obtain the number of cycles to failure.[51,52] With an estimate of the number of loading cycles that may be sustained during stable crack growth before failure occurs, it is possible to project component life.

The calculated number of cycles may or may not be adequate. Nevertheless, these techniques offer an approach for obtaining indications of component life based upon measureable quantities. If a longer lifetime interval is needed, tradeoffs can be made. For example, alternative operating conditions may be evaluated for decreasing stress level on the component. This will increase critical crack size at the time of failure. Decreasing the stress range will diminish the rate of crack growth, resulting in an increase in the number of cycles required, as this affects K. Using tougher material (i.e., higher K_{Ic}) will also increase critical crack size at failure. A more sensitive NDE method for confidently detecting smaller starting flaws would be another option. The acceptable crack-growth interval will also depend upon inspectability considerations. These relate to questions of inspection access, component geometry, and location.

In actual practice, these analytical procedures are more complex.[53] Stress-intensity factors in actual materials may be different from handbook values for reasons related to material heterogeneity, anisotropy, and other factors. Section size will also have an effect. Actual flaw geometry will differ from idealized forms used in textbook equations. Component loads are not so predictable and actual loading cycles may have configurations that do not resemble those used in determining experimental data. Conditions at the crack tip may not be entirely elastic but may involve varying degrees of plasticity. If so, assumed K_c values may not be valid. Then there are environmental effects that can accelerate crack growth.

To accommodate these variables and to take a conservative approach, larger initial flaw sizes may be assumed or smaller final sizes. It is often necessary to develop laboratory crack-growth data on the material to be used and to run the tests under simulated or actual service environments. Additional tests on actual hardware at various assembly scales may also be needed. NDE evaluations throughout the tests will probably be required to verify inspectability and flaw detectability assumptions.[54]

Sustained loading. This section, so far, has discussed design considerations for avoiding failures from cracks propagated under cyclic stress conditions. However, sustained loading can also cause initiation and propagation of cracks in some materials and environments. See Chap. 3 discussions on stress-corrosion cracking.

Crack initiation and propagation under these conditions were thought to involve crack-tip stress-intensity considerations similar to those governing crack propagation during brittle fracture and cyclic loading. That is, that the stress-intensity factor K might be a useful parameter for determining response of materials under stress-corrosion conditions as well. Experiments have confirmed this.

Tests conducted on precracked specimens under decreasing loads and within susceptible environments have shown that time to failure increases with decreasing loading stress. They have also shown that, below some lower stress value, failure does not occur. When the stress-intensity factor K for the test conditions is plotted against time to failure, as in an S/N graph of fatigue test results, the time to failure increases as K decreases. (Zero time corresponds to K_{Ic}.) The behavior is such that the lower portion of the curve becomes asymptotic to some K value, with no failures occurring below this K level, no matter how long the exposure time under stress.

The highest value of K that can be sustained and in which the starting precrack does not extend has been designated as K_{ISCC} and appears to be a material property. Consequently, this value of K for a given material–environment combination represents the upper limit of stress-intensity factor for avoiding stress-corrosion failures within the given environment.[55]

K_{ISCC} tests are usually conducted using specimens having one of two types of loading: constant displacement (wedge-opening loading type, or WOL) and constant load (cantilever beam). Values of K for the several specimen geometries and loading configurations may be calculated from equations that have been developed for these types of tests. Choice of test type is determined by available testing facilities and test material requirements. The WOL test can establish K_{ISCC} using a single test specimen, if necessary, and does not require extensive instrumented testing equipment. This test, therefore, is favored.

Identical values have been reported for all testing methods for a given material–environment system or combination, although material anisotropy can influence test results, making test specimen orientation an important consideration for specific design applications. For effective failure avoidance, K_{ISCC} values should not be exceeded in the vicinity of flaws and cracks that may be assumed to be present, even if below the NDE detectability limit.

Where stress-corrosion is the failure mode, the objective is to avoid crack initiation, because propagation to K_{Ic} under these conditions can be rapid, as has been noted. Crack initiation is best avoided through designing to maintain conditions below K_{ISCC}.

Fail-safe and safe-life concepts

Fail-safe design. Fail-safe terminology probably had its origin in attempts to ensure that failures of components did not injure people or cause other harm or damage to the equipment itself or to the surroundings. This terminology refers to two general types of approaches and both are described. For the first type, Hammer describes three categories of fail-safe design as (a) fail-passive, (b) fail-active, and (c) fail-operational.[56]

In fail-passive modes, a sensing of hazardous conditions results in deactivation of the hazardous agency or source. Fuses, circuit breakers, fusible links, and shear pins are examples. In fail-active systems, the system remains energized, but impending danger or hazardous condition is signaled through activation of a secondary system. Examples are alarms or indicator reactions to loss of integrity of safety monitors, such as periodic "beeps" emitted by battery-operated smoke detectors when battery capacity is becoming depleted and its reliability of operation is becoming marginal. Fail-operational sensors ensure that a failure does not trigger a worse condition. Interlocks that fail in an open-circuit position and gas valves that stop flow when they malfunction are examples.

Other types of fail-safe systems include "off" flags that are activated when aircraft navigation instrument displays fail because of malfunction, are turned off or otherwise inoperative, or cannot receive incoming signals. Centrifugal devices are installed on rotating machinery to detect overspeed conditions and signal the situation or shut down the system. Failure of some aircraft landing gear hydraulic systems releases a retention device permitting airstream and gravity to deploy and lock the wheels into their landing configuration.

These kinds of approaches do not lend themselves to all equipment-related hazards; however, when they do, they can be an effective means for forestalling serious failures. Reviews of hazards analysis and FMEA study results should be made to identify candidate components and to determine whether such concepts can be incorporated into the design. When these devices and techniques are employed, it is important to periodically verify that the fail-safe systems, themselves, remain operational.

In recent years, the fail-safe concept and terminology have been extended to include designs that maintain their integrity in the presence of ordinarily equipment-life-threatening failure mechanisms. In most traditional designs, failure of a load-bearing member, or progressive cracking, would tend to increase the load upon remaining members. This often hastens complete failure as high stress levels not only increase the stress-intensity factor K but can cause tensile overload. Fail-safe design must then provide sufficient residual strength, stiffness, and other required properties in the remaining members for continued operation. The operation might continue at normal or reduced loading until the failed member can be replaced or repaired, or until the equipment or system can be safely shut down.

The concept does not contemplate unabated and indefinite continued operation in the presence of a failed component—only the ability to bring the system safely to a standstill to afford an opportunity to correct the problem. *Safety* here refers to people, the equipment, and the surroundings.

A principal element in this concept is early identification of the failure. This requires inspection at appropriate intervals, adequate maintenance, and, for some systems, monitoring of pertinent conditions, properties, or events. Components that are inaccessible for adequate periodic inspection are not candidates for fail-safe design. Critical regions of such components should be designed for decreased stress levels or features that provide increased strength.

These practices should also be coordinated with life-extension assessments and activities. There is little to gain and much to lose in attempting to push a system or component to operate beyond its usable life. This is definitely not the province for applying fail-safe techniques. They are intended to accommodate those incidents that occur despite all reasonable precautions to avoid them and for critical applications where consequences of complete failure are particularly severe.

Another important element is the suitability of the system or component for application of fail-safe concepts. They are difficult and impracticable to apply to existing components and are, primarily, *design* techniques. The component design must be amenable to accommodating the increased demands that result from failure of one member. For mechanical components, this capability is expressed in terms of residual strength or the ability to carry a load in the presence of cracks. Cracked members cannot sustain their intended loads, requiring them to be shared by remaining intact members. This is a form of redundancy, although that term might more logically be reserved for situations involving replications of components or provision of *spares* or standby units.[57]

Development of fail-safe methods requires exceptional understanding of the hazards, damage mechanisms, and their principles and limitations. The roles of tensile stress magnitude, loading configuration, and operating environment in propagating cracks will dictate how design may be achieved in providing adequate residual strength.

One fail-safe design approach involves use of multiple-load paths. This requires stress analysis using computerized or prototype models to ensure adequacy and effectiveness of this approach. Crack arrestors are another fail-safe element that is used in the aircraft industry to confine the extent of running cracks.

Crack arrestors may take several forms. One involves riveting together several individual components (in flat panel assemblies, for example) with the joint constituting a crack interruptor. Surface-bonded doublers similarly interrupt the progression of cracks and confine them to regions between doublers. In welded monolithic structures, use of materials having superior fracture toughness properties welded into the structure in critical regions are often used for this purpose. If sufficiently tough and wide enough, such regions will cause propagating cracks entering them to blunt-out and stop.[58–60]

Safe-life approach. *Safe-life* generally refers to damage-tolerant fatigue crack avoidance techniques described earlier in this chapter. The usefulness and dependability of this approach require accurate characterization of operating conditions, including well-defined maximum applied and mean stress levels, cyclic-loading configurations, and the service environment. Geometric features and other sites of stress concentration and the presence of residual stresses must also be well characterized.

Crack-growth characteristics of the material under the expected and reasonably anticipated service conditions should also be thoroughly understood. And there must be a high level of confidence in the NDE methods used for establishing the minimum detectable crack size. A safety factor of at least two is often recommended in this respect. That is, if the minimum crack length that is believed to be capable of being confidently detected every time is 5 mm (0.2 in), then the safe-life design should assume 10 mm (0.4 in).

Using the best available data for these parameters and conditions, a usable lifetime for the component is projected, as described earlier in this chapter. If a reasonable operating interval is projected, the component is placed into service, then removed when the predicted lifespan has elapsed. Presumably, if calculations are correct, the data are valid and accurate, and operating conditions closely duplicate those assumed during the design stage, the projected safe-life interval will have been free of fatigue failure.

The service interval may be selected according to the level of the designer's confidence in the test data, criticality of the application, experience with similar components in similar applications and environments, and predictability of the operating situation over the projected lifetime of the component.[61,62]

Scope of application

The design concepts and techniques described here have been developed and used to date mostly for aircraft, aerospace, and military applications. Requirements where these strategies are most important involve considerations of maximum attainable strength and minimal weight and where costs of achieving these objectives is not a dominant factor. Obviously, these methodologies are not for everyone and every application. For most applications where fatigue cracking is a concern and which are not subject to rigid performance criteria, less rigorous approaches would probably be more practicable and cost-effective.

Notwithstanding this, the principles used in these design approaches are well founded and have been tested in demanding real-life applications and, for the most part, have been successful. Therefore, they deserve the attention of every design engineer, whether the applications being pursued warrant the full rigorous analysis of these techniques or not. Their basic principles will work on any design and working knowledge of them can be an asset.

The need for these approaches and other analytical techniques increases with increasing probability of brittle fracture. This is a consequence of more intricate and dynamic loading and complex designs, particularly those employing computerized design database systems, which tend to result in leaner structures and minimal safety factors. Other adverse influences are the need for higher stress, use of higher performance materials, and thicker sections. The need for more aggressive environments, including higher temperatures and pressures and shorter cycles, further intensifies the likelihood and propensity for material damage. And the trend toward the fabrication of monolithic structures through increased use of welding worsens the consequences of a propagating brittle crack.

Although all elements of a formal fracture control plan developed for the design and fabrication of aerospace components may not be wholly applicable to more mundane components, these approaches constitute ideal models for evaluating design strategies and procedures for any application. It appears inevitable that, in time, these philosophies and concepts, if not the methods themselves, will be extended to many other systems. As the techniques become more widely used and understood, and as needs for lighter weight and stronger structures and components increase in many industries, the extension of these techniques into these other areas will also increase. It is prudent to be well prepared for such developments.

Design philosophy

As noted earlier, there are numerous failure mechanisms that the designer must be aware of and avoid (see Table 3.1). Because of time and space limits, we have concentrated upon the most prevalent failure mechanisms and those that are most damaging and difficult to avoid or control. It is important, therefore, that the designer understands that *all* failure modes must be avoided in the component, system, or equipment being contemplated. It would serve no purpose to design a component that might never fail by brittle fracture or fatigue and yet fail by buckling, corrosion, or excessive wear. Also, it should be appreciated that designing against brittle fracture, for example, cannot be accomplished through confining attention solely upon the material. All the factors influencing it must be considered, such as section thickness, operating temperature, flaw detection methods, and enviromental variables. These same considerations also apply to other failure modes.

There are many failure modes, many more combinations of conditions, and a virtually unlimited number of combinations of applications, materials, failure modes, and people interactions that can cause failures of engineered products and systems. Each design situation is, therefore, unique and must be evaluated on its own, requiring its own approach to avoid failures. This is why the use of analytical tools has been emphasized in both the assessment of potential problems and in the application of techniques and strategies to avoid them. It is just too much to attempt to handle intuitively.

For mechanical systems and components, these analyses usually identify fracturing failures as the most critical. These are materials dependent and avoidance of these problems centers about material properties and how the materials are used. General guidelines for safe design practice must evolve from knowledge of factors that influence crack initiation and growth. These include, but are not confined to, the level of stress imposed upon the part, including residual stresses and stress concentrating effects; the material of construction, its properties and microstructural characteristics; the service environment; part thickness, geometry, and general configuration; characteristics and sizes of existing cracks and flaws; fabrication processes and their thermal and mechanical effects; and inspectability.

Material substitutions and other design decisions must be made carefully. For example, higher-strength material than necessary may not provide sought-after safety margins. This is because of usually lower accompanying fracture toughness and higher ductile–brittle transition temperature. Complex designs and those requiring a multiplicity of welded joints can produce high levels of residual stress and distortion. Simple designs are usually preferable and can avoid many problems and adverse material conditions.

Economy grades of material may not be cost-effective in the long run when their needs for intricate crack size analysis and monitoring (because of marginal K_{Ic} fracture toughness values) are compared with the superior properties of better-quality steels having higher K_{Ic} values and correspondingly larger critical crack sizes. This will allow use of less-sensitive NDE methods and, overall, may result in significant cost savings.

In other words, money spent on cleaner steels produced by fine-grained aluminum-killed practices or other grain-refining deoxidation techniques and, thereby, offering superior toughness properties can save the often greater expense of extensive testing and inspection to verify product integrity. Purchase documents should be explicit, however, in requesting the desired properties and minimum acceptable values for K_{Ic}, nil-ductility temperature (NDT), or other fracture toughness criteria.

"Adequate" fracture toughness must be determined through evaluations against design requirements and many other factors. Fracture toughness is often expressed in terms of a K_{Ic}/(yield strength) (i.e., static load) or K_{Id}/(yield strength)$_d$ (i.e., dynamic load) ratio. Values of 1.0 or better are usually considered desirable, although materials having values below this are often used, either out of necessity or for optimal response, considering all factors involved, and under the proper design and service conditions.

When evaluating fracture characteristics and considering approaches for improvement, it is advisable to address all three factors that generally contribute. These are the level of tensile stress on the component (including residual stresses), size and characteristics of flaws, and inherent toughness of the material at the service temperature, loading rate, and thickness being used. Brittle fracture may be avoided through improvements in any or all of these factors.

Rolfe and Barsom provide an excellent summary of effects of these factors.[63] They point out that all three can have a major effect upon component life. For example, the crack-growth rate decreases significantly as the applied stress range is decreased. Structures loaded slowly permit K_{Ic} to control, whereas, under impact or high-loading-rate conditions, the lower K_{Id} controls. Also, fabrication and inspection quality determine flaw size, and fatigue crack-growth rate for small flaws is low. Use of material that exhibits elastic-plastic behavior represents a major improvement over that exhibiting plane-strain behavior.

Wherever possible, materials used should have sufficient notch toughness to preclude failure by brittle fracture under even the most severe operating conditions. Critical designs should incorporate fail-safe features (e.g., multiple-load paths or crack arrestors). Sharp notches or surface grooves, particularly running normal to the direction of principal stress, should be avoided. Similarly, sharp edges should be rounded and section thickness or geometry changes should be gradual transitions with large-radius fillets. Surfaces subject to corrosion should be suitably protected. Risks of hydrogen embrittlement from electroplating of susceptible materials should be avoided through use of substitute (nonelectrolytic) plating methods or other coating systems.

Specific guidelines for actual components and applications should be obtained through reviews of analytical studies, such as hazards analysis, "what if," FMEA, HAZOPS, and FTA. The design staff should obtain the services of specialists, as needed, in dealing with issues outside their fields of expertise. This is especially desirable in considering specialized tests to develop design data or to evaluate the response of materials to various loading configurations and environments. Hazards involving other disciplines and their avoidance should also be evaluated by those having expertise and experience in those fields.

Minimizing Product Flaws

Transition from design to production

The route that leads from design concepts, drawings, and specifications through engineering and prototype evaluation to production is not an easy road. Design reviews at appropriate stages with participation by representatives from all departments involved in the product can relieve many potential difficulties. One of the primary values of design reviews is the establishment of personal contacts and interdepartmental communication channels for addressing and resolving the inevitable problems and discrepancies that will arise.

The manufacturing stage is responsible for converting designs and specifications into tangible working machines and equipment that, in all respects, fulfill designers' intentions. The finished product (subassembly or system) will consist of materials and components that were obtained elsewhere and incorporated into the product. Some will have been cut or shaped and fabricated into different forms, some will have been joined to other fabricated parts or purchased components, and various mechanical, structural, and electrical interconnections made. User interfaces, controls, enclosures, protective coatings, labeling and packaging, plus installation and maintenance manuals and warranty information complete the product "package." It may be big or small, simple or complex, inexpensive or costly, mass-marketed or custom-made. Each of the possibly large number of steps required in its fabrication, assembly, and completion are opportunities for deviation and for introduction of discrepancies, errors, and hazards-producing flaws and defects.

Therefore, it is essential to the preservation of safety features of the product that everyone involved is knowledgeable concerning the need for keeping defects out of the product and the precautions and techniques for doing so. It does no good for a lone "safety champion" in the organization to promote safe product design while others remain unsympathetic to these issues or are unconvinced of their need and value. Under these conditions, the consensus usually reverts to manufacturing the product according to established traditions. This is why safe product design must be a corporate-wide policy solidly backed by upper management.

In considering the subject of product flaws or defects, it is well to remember that this covers virtually every industrial activity. Courts tend to define *defect* according to the definition given by a products litigation attorney, mentioned earlier, as anything wrong with the product that may have a causal connection with harm, injury, or damage. The law does not yet require a manufacturer to guarantee that its products will not injure someone, but it does require a guarantee that they are not defective (by the court's definition).

Products can be anything manufactured and sold. They may be solid, liquid, or gas. Besides the physical product itself, courts have considered the following to be *products* and, therefore, subject to the laws of products liability: packaging, labeling, warranties, use and maintenance manuals and instruction sheets, installation and service, promotional and advertising materials, datasheets and catalog listings, and replacement parts and components. A proposed Uniform State Product Liability Act has defined *product* as "any tangible object or good including any service provided in connection therewith."[64] This definition and other provisions of the model act may not have been adopted by all jurisdictions; however, they reflect the thinking of litigators. It could be a mistake for a manufacturer to insist upon defining *product* by a more restrictive standard.

Therefore, the task of the manufacturer in keeping defects out of its products encompasses virtually every activity in the company. Manufactured products produced in large quantities and marketed on a wide scale, as is the objective of most manufacturers, require special care, as a feature that has propensity for harm or injury is an incurred risk that is multiplied by the number of products sold. Not everyone purchasing or using hazardous products will be harmed and not everyone harmed will pursue legal action, but only one damaging lawsuit can have serious economic consequences for a manufacturer.

In evaluating the range of product flaws and defects that can incur liability for a manufacturer (and others in the product distribution chain), Fig. 4.1 is a short list that summarizes some of the more common categories that have been a source of trouble. Some of these are related to design, but practically all can be introduced or affected in some

- Materials used or substituted that are of inferior, substandard, or questionable quality
- Noncompliance with or indifference to mandatory safety regulations or voluntary industry standards, codes, and state-of-the-art
- Incompatibility with intended and probable use and foreseeable misuse and abuse
- Inadequate warnings of potential harm from unavoidable hazards
- Unanticipated degradation and adverse response
- Unjustifiable reliance upon quality of materials and components supplied by others
- Preventable risks that outweigh product usefulness, social benefit, and cost of elimination
- Personnel competence inconsistent with product complexity and performance
- Product performance, safety, reliability, and fitness for intended service inconsistent with promotional claims

Figure 4.1 Product characteristics and manufacturer conduct that can incur liability.

way during manufacture or production. It is well to note that in the vast majority of cases, these and other factors that can lead to liability are the result of well-intentioned actions or innocent omissions. Deliberate violations of safety practices or willful introduction of hazardous features are usually not the issue.

Engineering

Before the product can be considered for full-scale production and after the design is well-defined, various questions related to transforming the design into production items must be settled, including evaluation of prototypes. These steps may be considered a design activity, but there are reasons why others should handle it. One is to ensure that the product or system is doable with available facilities and within budget and schedule, and, if not, what must be obtained or changed to accomplish it. Second, to save time and provide effective feedback to the design function, it can be advantageous for some portions of engineering and prototype work to proceed concurrently with design.

During this transformation, generally termed *engineering,* a number of issues are addressed. A principal item is production feasibility. Modifications may be required to match facilities and capability and to provide necessary tolerances inherent in production operations to avoid potential conflicts and stoppages. This step verifies that all specified parts, materials, and components will be available and are consistent with processing capacities and capabilities. And it confirms that procedures and devices intended to provide the required degree of safety will, in fact, be adequate. In some cases, they will prove excessive and be found to interfere with the product's functioning and require revision.

As is often discovered, the whole can be much less than the sum of its parts, and projected performance of the completed product or system may fail to live up to what would have been expected based upon performance of individual components. A few moments spent in evaluating a real product can provide more information than days of paper calculations and computer modeling. Although there is a time and place for such calculations, there is also a time and place for evaluating the real thing, as there is no effective substitute for actual operating experience.

If at all possible, a working prototype should be constructed using the design drawings and layouts and commercially available materials and components. It should be constructed using working tolerances and shop practices closely simulating those of actual production. All operations should duplicate as closely as possible those that will be used in full production.

These tests should be formally structured and planned, fully documented and recorded, and cover all operational aspects and potential hazards disclosed during FMEA, FTA, and other studies. Tests should be conducted in typical as well as nontypical but foreseeable use environments. Safety issues deserve special attention and these should be evaluated by typical users, not by trained product development engineers familiar with the product. Data from instrumented runs should be compared with design projections to confirm expected response or to identify and evaluate discrepancies. The extent and nature of the tests depend upon the product, equipment, or system and the judgment of the engineering staff. A temptation to foreshorten this phase and to elect to rely instead upon early customer feedback from initial production runs should be stifled, as such a move can defeat the purpose of the failure avoidance program.

Tests conducted on a single product prototype may not be as relevant as when several items are evaluated in the same series of tests. While each item may be instrumented and monitored for specific behavior or response, many characteristics are not measurable but must be evaluated subjectively through personal observation. Such evaluations lack comparison standards and provide no firm basis for judgment decisions. Therefore, it is desirable that, wherever feasible, each product test include reference standards that are evaluated at the same time and under the same conditions as the product being evaluated. The reference product or control sample might be a previous model for which field experience is available, or a competitive product.

Positive results from in-house tests may suggest the need for controlled field tests or extended prototype evaluations. The purpose of testing production prototypes is to determine if the product or equipment satisfies design objectives and that it does not present hazards from its existence, storage, typical use, foreseeable misuse and abuse, and disposal. Some of these characteristics may not be discernible during evaluations conducted under laboratory testing environments; therefore, tests should be arranged to be conducted under conditions sufficiently representative of actual field use to provide these indications. Guidance may be obtained in this through reviewing failure studies of similar products or systems or results of design analyses where critical operating variables or failure susceptibilities are identified.

Field evaluations of prototypes are useful in that they provide real-world experience and response. A disadvantage lies in the difficulty of control and in obtaining feedback information. Some provision should be made for recording operational information and response to critical factors. In addition to this, user or operator opinions and impressions should be obtained in detail using specially prepared forms and questionnaires. After a prescribed test period, the product or products should be retrieved and returned to the manufacturer for disassembly and evaluation. This is essential in the event of a malfunction or failure, particularly if there is an injury or damage associated with the failure.

The test procedures used during this phase depend largely upon the nature of the device, product, or system involved. The need for feedback from the field never ceases, even for products that appear to be well established and successful. Despite in-plant quality controls and inspections, problems can creep into a mature product as raw materials, components, and suppliers change, as deterioration during exposure to use environments occurs, and as manufacturing processes and fabrication tolerances vary. A definite procedure and schedule should be provided for obtaining this information, and it can take many forms, including spot recalls, service inspections, customer complaints, and reports from field representatives.

Procurement

Product flaws and defects can originate during any step of the production process, starting with purchase and procurement. The designer's specifications must be converted into tangible materials and purchased parts that will satisfy the intended purposes. This is a likely place for deviations to occur. There will be the inevitable pressures from suppliers to substitute "just as good" and "equivalent" grades of material for those specified. Unfortunately, common grades that are usually in abundant supply for immediate delivery do not always conform to performance criteria or quality standards required for safe product designs (possess the desired level of fracture toughness, for example). Grade stamping or certification to some American Iron and Steel Institute (AISI) or ASTM code or practice does not necessarily ensure conformance to the particular requirement specified by the designer.

Characteristics that are important in purchased materials for maintaining product integrity in fracture-safe design include strength and fracture toughness properties. These should be expressed in terms of relevant service conditions and material orientations of interest (i.e., longitudinal, transverse, or short transverse). Chemical compositions, microstructure, and heat-treated conditions are other important factors. It is advisable to request copies of mill certification documenting test response for the specific lots or heats of material being shipped to ensure traceability.

Purchasing personnel usually are unfamiliar with the engineering significance of apparently minor compositional differences and mechanical or material property variations. They may fail to appreciate why seemingly satisfactory material that is immediately available from local stock at a good price is unacceptable, while the desired grade is available only on special order and at extra cost. There will be pressure to accept the immediately available material, and this is where problems can originate. This does not mean that concessions should not be made, as they are sometimes unavoidable. However, the designer or user must be knowledgeable about material grades and their influence upon the product as designed and intended. Minor compositional differences that may be irrelevant for the application or dimensional tolerance deviations in plate thickness that often disqualify a lot of material from grade certification may not affect its usability for specific applications.

Along these lines, FMEA and other materials-related analytical studies should be consulted during procurement negotiations to identify critical performance requirements for specified items and components. This is necessary to ensure that appropriate materials are obtained. This is especially important for certain categories of items for which there are numerous grades of varying quality and suitability—fasteners, for example.[65–68]

Production

Manufacturing operations. Manufacturing typically is the stage devoted to converting incoming parts, materials, and components into functioning products, equipment, or systems, and doing so repetitively. Depending upon the product being manufactured, the resulting items may or may not be identical, and there may or may not be large numbers produced. Nevertheless, the operations being performed on each are generally similar, as this operational repetition and stepwise progression from one manufacturing station to another are efficient and economical.

Manufacturing has been classified in various general ways. One scheme has seven categories: (a) casting or molding, (b)cutting and forming, (c) material removal or machining, (d) heat treatment, (e) finishing, (f) assembly, and (g) inspection.[69] Processes that involve material removal, cutting and forming, and some assembly operations are subject to inspection verification and basic dimensional checks. Deviations originating in these processes, when adequately inspected, do not contribute to a great extent to product failures as they are evident using basic techniques or show up during subsequent assembly stages.

Quality or adequacy of casting, heat treatment, some finishing and assembly operations, and joining processes are more difficult to assess. Visual appearance and dimensions are not useful criteria for evaluating quality characteristics for material processed during these operations. Also, most of these operations tend to be either performed manually or are manually controlled and their results evaluated subjectively or indirectly. These manufacturing processes, therefore, can be a source of product defects and require special attention.

Metal casting. Casting is, essentially, production of material and component form in a single operation. Constituent raw materials or melt stock master alloys are melted and metallurgically treated to eliminate impurities, then poured into molds to provide semifinished parts. Sometimes they require some machining, finishing, and sizing operations, but usually they are placed into service in their as-cast condition.

Properties of the cast components, as well as their soundness and integrity, are functions of the quality and purity of starting materials, alloying additives, melting practice and melting environment, process used, pouring temperature, mold design, and other factors. Production of quality castings requires vigilance and control over each of these factors and should be specified and closely followed by shop floor personnel.

Inspection standards must be developed and implemented to ensure that each lot or heat meets the requirements. Heat-to-heat variability in properties and other characteristics for supposedly identical castings is not uncommon and can arise from a number of sources, such as those mentioned above. Castings are subject to compositional variability and heterogeneity such that the microstructure and mechanical properties, including weldability, can differ from one end of a casting to the other.

Casting process control does not usually lend itself to instrumented techniques, with the result that quality is often heavily dependent upon the skill and experience of foundry personnel. The inherent variability associated with castings should be appreciated by designers and quality control personnel to ensure that cast components are not defective and a cause of failures.

Powder metallurgy. Fabrication of small parts from metal and alloy powders is an increasingly attractive option for many components as an alternative to casting. By this process, mixtures of elemental and prealloyed powders blended with binders and lubricants are compacted under pressure within suitably configured dies. The resulting "green" parts have sufficient strength for ejection from the compaction dies and limited handling for transfer to baking ovens for removal of the binders and lubricants. They are then placed within protective atmosphere sintering furnaces where the powdered constituents diffuse and result in a semidense solid of near-net shape (of 90+% of theoretical density).

For applications requiring full density, supplemental compaction methods such as hot isostatic pressing (HIP), closed-die forging, or liquid-metal infiltration are capable of producing full-density solid material. These processes are cost-effective and efficient and are being used in the manufacture of an increasingly broad range of parts and components for applications ranging from household and consumer products to automotive engine and transmission parts, and even turbine disks for aircraft engines. The latter can have properties superior to cast-and-forged components. Like castings, however, part quality, properties, and performance are a result of the proper starting materials, proper blending and compaction, and proper control of all subsequent process steps. Here again, part quality and suitability depend almost entirely upon process control and condition monitoring.

Metal joining. As discussed in Chap. 3, welding and other related joining processes can also be a potential source of problems, and for most of the same reasons as casting, plus some additional reasons. Welding, brazing, and soldering are prone to difficulties because of the large number of process variables involved. Automated techniques can significantly improve the situation, provided they have been developed adequately with proper control tolerances.

Weld quality is affected by base metal composition as well as by filler metal composition and welding conditions, although some welding processes and joint configurations do not require filler metal addition. Base metal weldability often is affected by the presence of trace elements, and their tolerance thresholds can depend upon section thickness and joint restraint, weld cooling rate, weld travel speed, weld bead size, and other factors.

The apparent ease and simplicity of many welding and other joining process variations create opportunities for defects. It is often another example of the technical sophistication of the tool outstripping the understanding of its user. Partial-penetration welds (fillet welds, for example) are notorious for originating fatigue cracks as are the normal weld bead profiles and rippled surface features. Unfused weld bead roots constituting a sharp notch or crevice should not be stressed in tension, particularly under cyclic-loading conditions.

Whenever possible, especially for critical components, welds should be located away from corners, regions of stress concentration or where part geometry changes abruptly, and away from the highest stressed and highest temperature regions. Weld metal properties sometimes listed in handbook tables may not be reliable as guidance because the grain-coarsened heat-affected zone may be limiting for the material. Weld properties representing across-the-joint tests (including the base metal, heat-affected zone, and weld metal) are more relevant.

Depending upon the application, it may be necessary or desirable to heat-treat completed welds to develop optimal properties. For some alloys and heavy sections subject to cracking or brittleness due to rapid cooling or quenching, it may be essential to preheat before welding and maintain a minimum interpass temperature and slowly cool to ambient conditions. Weld heat treatments may be complex and require controlled heating and cooling rates, holding intervals, and quenching steps followed by reheating, holding periods, and cooling. Each of these steps must be controlled to within the prescribed limits as they can have significant effects upon material properties, particularly upon strength level, fracture toughness, and ductility.

Brazing and soldering are metals joining processes wherein the base metal generally is not melted into the joint, although some very superficial alloying takes place at the interface to effect wetting of the filler metal to the surface of the base metal. Maintaining of adequate clearances between mating surfaces being brazed or soldered is important in developing optimal joint properties, and this may be controlled through machining of proper joint configurations. Brazing and soldering production operations are usually automated, which improves the statistical reliability of processed parts. Preplaced fillers followed by heating within controlled atmosphere furnaces and ovens provide production efficiencies and cost-effective batch or continuous-process operations.

For metals joining operations involving materials that may be subject to defects and other problems, their weldability should be established for the anticipated production conditions through a series of preproduction welding or other joining tests. The same process conditions should be used as for the production operations and the resulting sample joints examined structurally and metallurgically for suitability. The application may require fracture toughness determinations for the weld and heat-affected regions, stress-rupture, stress-corrosion, or fatigue crack propagation tests.

Weld data developed for one set of welding conditions may not be transferable to another. In the author's experience, a nickel alloy formed into continuous thin-walled small-diameter tubing from a flat strip was weldable with resulting high-quality welds when welding was performed at moderate travel speeds. When welding speed was increased to the desired production rate, centerline liquation cracking occurred in some lots. It was found that threshold tolerance for harmful trace impurities and the need for corrective treatment during melting of the alloy were dependent upon the weld travel speed or production rate. This example demonstrates the need for duplicating production conditions in any manufacturing weldability evaluations or weldability acceptance testing that might be conducted on incoming material.

As with other failure-related material characteristics, weldability tends to decrease with increasing strength levels, among other properties. Where welding of high-performance materials is required during production, special care must be exercised during preproduction testing to establish weldability and process conditions to ensure that part restraint and all other characteristics of the actual component are closely duplicated. Where weld performance is a major factor in the integrity of the structure or part, mechanical property data should be developed specifically for the welded joint, as data for the unwelded plate of the same alloy may give overly optimistic indications. This holds for all properties and behavior, including fracture toughness, stress-corrosion resistance, fatigue cracking, high- and low-temperature response, general corrosion and erosion, and others.

Since most welding, brazing, and soldering operations involve localized or general heating to high temperatures, their side effects can damage the part itself or surrounding areas and lead to failures. Heat-sensitive materials, particularly organics, must be shielded or protected from these operations. In addition to heat, welding arcs emit ultraviolet and infrared radiation, ozone, and other gases, fumes, and particulates that can deteriorate or adversely affect other components and materials.

Weld spatter (small molten metal droplets expelled from the weld region in some processes) can adhere to adjacent surfaces. Accidental arc strikes on hardenable steels have been identified as the origin of catastrophic brittle fractures. Some welding, brazing, and soldering processes employ fluxes to assist in cleaning the region to be joined of oxides and other surface films and contaminants. These fluxing substances are aggressive at joining temperatures. Their residues can lead to deterioration and corrosion at ambient temperatures. Residues of welding slags have triggered accelerated attack of nickel alloys in low-level sulfur-containing petrochemical process atmospheres as a result of embrittlement caused by their sulfur-scavenging properties at the operating temperatures involved. It is good operating practice to remove all traces of fluxes or slags from welded components thoroughly.

Heat treatments. Heat treating, baking, or drying operations can be critical to the successful performance of manufactured parts and, as for other processes being discussed, visual inspection is an inadequate method of evaluation. Heat treatment of metal parts often affects mechanical properties and in some cases provides a basis for evaluation of the process and its adequacy. The metallurgical condition is only one result of heat treatment that can affect the integrity of a metal part. The heat treatment can cause distortion if phase transformation is involved or if there are residual stresses in the part or structure beforehand (from welding or forming, for example). Also, distortion and residual stress can be introduced during heat treatment as a result of the geometric configuration of the part. Such stresses can add to service stresses and lead to higher fatigue crack rates or aggravation of other stress-related failure mechanism.

Drying and baking operations can be important in processes that are degraded by preexisting moisture. Often these operations are elements of another process or procedure. These steps usually must be performed within a certain time of a subsequent operation to ensure that the part does not reabsorb moisture after the baking or drying treatment. They can be critical in avoiding brittle fracture due to hydrogen embrittlement in electroplated parts. Time, temperature, and the interval between plating and baking are important variables.

Coating processes. Coating processes can affect the corrosion resistance of metal parts and must be properly carried out, including adequate preparation and monitoring of the process. Coating adhesion, soundness, thickness and uniformity, surface appearance, properties, and effectiveness of coating protection are influenced by control of process parameters.

A number of processes are used and include electrolytic, electroless, electrostatic, vapor-deposited, mechanical, and thermal. They are applied to impart some surface quality and the outcome, results, or performance depend upon adequate control of process parameters.

Tests and inspections. Many activities considered to be manufacturing are often assembly operations, where components manufactured by others are assembled into larger components, subassemblies, or the final product. Depending upon the product being produced and the nature of its components, there should be some kind of in-house verification, acceptance testing, or evaluation of these items before their use to ensure that they are suitable.

Verification tests should evaluate critical factors as noted in FMEA lists. The testing method should be sufficiently discriminating and relevant to actual use to reveal critical conditions essential to safe operation. It should include all parts that are vulnerable to failure or malfunction and that could be implicated in personal injury or harm. Faulty or defective items should be clearly identified and well separated from acceptable ones and either repaired and retested, discarded, or returned to the supplier for repair or scrapping. Steps should be taken to ensure that rejects cannot be intermixed with usable parts anywhere in the plant.

All test equipment throughout the plant should be maintained in good working order and calibrated at prescribed intervals in accordance with manufacturers' or industry standards. Lack of vigilance in this respect has led to liability when defective components were not detected during inspection and caused injury and damage.

Whenever the reject rate is high or inspections reveal a consistent pattern of problems or constant deviation from specified values or performance, engineering and design personnel should be advised, as this may be symptomatic of design or specification deficiencies. These kinds of difficulties normally would have been discovered during first-article, sample, or prototype evaluation.

Lack of relevant regulatory requirements or voluntary industry-wide standards does not relieve a manufacturer from verifying the quality of components used in its products. If no standard exists, one should be developed in-house to ensure that defective materials or components do not get by and into the product and compromise safety or degrade performance.

Labeling and instructions

Ordinarily, and until the relatively recent developments of products-liability law and related regulations, a product's labels would have been considered the province of advertising and promotion. Since products cases have been decided solely upon what was or was not on a product label, instruction sheet, or product packaging, this can be an important issue to a manufacturer. The manufacturer has a legal obligation to provide instructions on safe and proper use of the product and to explain any nonobvious or hidden hazards or risks. It should also describe precautions for use under abnormal but foreseeable conditions, including a clear explanation of consequences of failing to heed the warnings, and corrective action to be taken.

The preparer of instructions and warnings must cover the entire life cycle of the product and not limit attention to usual and normal use. Abnormal situations that may be nontypical but foreseeable must be addressed. These include consequences of prolonged storage under adverse conditions, possible deterioration of the product or its container during such storage, results of accidental damage to the product or container, consequences of product spills and inadvertent contact with other materials and possible reactive effects, and the safe handling and disposal of unused product, residues, and its packaging or container.[70] Any hazards or release of toxic substances that may accompany failure to observe and comply with these warnings should also be explicitly and prominently explained, along with corrective or neutralizing action.

The extent of information provided with the product, its content, and form should be consistent with the characteristics of the product, its complexity, and consequences of misuse, malfunction, and other circumstances. For simple products with minimal hazards, a one-page sheet, tag, or label may be sufficient. Large complex systems may require a set of manuals that occupy an entire bookshelf and include subsystem flowcharts, fault identification networks, and even computerized database diagnostic and maintenance information and parts indexing. As noted in the preceding section, product warnings, labeling, and instructions are design functions; however, it is during production that they are implemented and made to accompany the actual product. It is important, then, that their value, content, and other factors are not lost or diminished between the design specification and its implementation.

Warranties

Product warranties have not been discussed so far, and this should not be construed as an indication of their insignificance. They can be most important. Since 1957 when the Uniform Commercial Code (UCC) was drafted, it has been adopted by every state and is now considered the uniform law governing sales in this country. Although it is a lengthy and exhaustive document, sections dealing with express and implied warranties are of primary interest within the context of our immediate discussion on defectiveness as related to product labeling, instructions, and liability issues.[71]

The UCC has established minimum standards of quality for virtually all goods sold in business transactions. Fitness of a product for *ordinary purposes* can be degraded by a condition that makes it hazardous to use or contributes to property damage or other harm. Under the Code, lack of fitness for ordinary purposes constitutes a breach of implied warranty and a basis for liability as a result of damage stemming from that breach.

Other sections of the UCC cover implied warranty of fitness for a *particular* purpose, where the seller had reason to know the user's specific purpose for the product and where the user or purchaser relied upon the judgment and experience of the seller in choosing the product to satisfy that purpose. The UCC contains provisions that permit the seller to disclaim (exclude) implied warranties from the transaction, provided that the seller makes it clear that the product is worn, used, or otherwise inferior in some way, through conspicuous written statements that the sale is conducted on an "as-is" basis.

The UCC was intended primarily for commercial transactions. The Magnuson–Moss Warranty Act of 1975 was enacted to improve the adequacy of information available to *consumers* (in contrast to commercial dealers) to prevent deception and improve competition in the marketing of consumer products. It does this, among other provisions, through providing minimum disclosure standards for written consumer product warranties.[72]

This legislation governs warranties and disclaimers for use with the sale of consumer products and, therefore, is less flexible than similar provisions of the UCC. This is especially so for disclaimers. The Magnuson–Moss Warranty Act provides for full and conspicuous disclosure, in simple and readily understood language, of terms and conditions of consumer product warranties. In addition to this, it specifies terms of sale and distinctions between *full* and *limited* warranties and their format, service contracts and insurance, remedies for settling disputes, and so on. The legislation does not require warranties on consumer products but imposes specific duties and liabilities upon those who offer them. Its primary aim is to eliminate the practice that is detrimental to consumer-products purchasers (and permitted by the UCC), of embedding fine-print disclaimers of implied warranties within the content of written warranties.[73]

The thrust of this warranty legislation, within the context of our discussion here, is to limit the use of disclaimers to eliminate implied warranties in the sale of consumer products. Manufacturers are motivated in this by the desire to avoid liability. While this is permitted, courts tend to disregard disclaimers that they view as unconscionable or in violation of public policy. In other words, disclaimers should not be regarded as substitutes for making safe products. Reliance upon disclaimers to relieve a seller's concerns over product liability within the *consumer* realm is risky business. Courts regard consumers, in contrast to commercial buyers and sellers, at a disadvantage with respect to technical understanding, product quality, and consequences of defectiveness; consequently, they usually side with the consumer.

Packaging

An initial reaction is to ignore product packaging. The temptation is to take the position that it is the contents that really matter and this is where the emphasis should be. Unfortunately, nothing is as simple as it appears to be, or was at one time, and packaging is a prime example.

Many lawsuits have been brought over some aspect of packaging and it deserves a manufacturer's close attention. In fact, packaging should receive the same caliber of design and failure avoidance scrutiny as the product being contained by it. This is sound practice because, in the eyes of the law, the package is a "product" all its own and is subject to the same legal tests of defectiveness and other characteristics as the product itself.[74]

Packaging has more functions than one might at first suspect. Its usual purpose is to contain, protect, and preserve the product during its transit from the manufacturer, through the distribution chain, and to the ultimate consumer or user. It is far from a direct route and a product and its components may accumulate thousands of miles over land, water, and air in the process. Often, the product remains within its package or container throughout its useful life and even beyond. The package can serve as a dispenser and measuring device. Packages also play a prominent role in sales inducement and customer motivation, and they provide a means of promotion and advertising.

A bewildering array of rules and regulations govern packaging. The list of major regulators includes the U.S. Postal Service, Interstate Commerce Commission (ICC), Food and Drug Administration, U.S. Department of Commerce, U.S. Department of Transportation, Federal Trade Commission, military services, fire departments, and many other municipalities and authorities. Some industries have developed their own consensus standards on packaging. ASTM lists well over a hundred standards for testing the properties of packages and packaging materials of paper, paperboard, containerboard, corrugated and solid fiberboard, plastics, glass, wood, and metal. The standards specify minimum requirements and the same criteria apply as in assessing compliance with product regulations. Compliance may not be a strong defense but lack of compliance with minimum standards can be damaging.

Packaging can be defective in various ways. Product spills on public roads, highways, city streets, and waterways are commonplace. Most observers view them as inconveniences, but when the spill involves toxic materials and extensive areas of contamination, the product loss becomes insignificant in comparison to costs of cleanup, traffic control, residential evacuations, and litigation resulting from harm and damage from contact with the material. Defective packaging is not always the cause of these incidents but it frequently is.[75]

Even more damaging can be a container failure that permits its contents to damage materials stored in adjacent areas, or to react explosively with other chemicals or substances and cause a catastrophe. Defective containers can allow their contents to be contaminated in various ways and become harmful or worthless. In recent years, incidents of product tampering have caused much concern by the public and manufacturers and packagers of the products involved.[76–78] These incidents have led to regulations for tamper-resistant and tamper-evident packaging for a broad spectrum of pharmaceutical and healthcare products.[79] Packaging that prevents opening and ingesting by children of various harmful household and over-the-counter products and prescription drugs is well known.

There should also be no reaction between the product and its container throughout its lifetime. Such a reaction can damage the product and its container and possibly degrade and weaken it and cause it to fail. The reaction may create internal pressure and lead to ruptures that could injure people.

Even after the product has been separated from its packaging, the packaging must be disposed of and, therefore, its risks of damage to someone or something still exist and must be considered. Usually, discarded packages contain product residues, and these may be hazardous on their own or may react with the packaging during exposure to the weather. Liability of manufacturers for damage sustained through contact with discarded containers depends upon many factors; however, plaintiff attorneys can usually readily identify the manufacturer from the name and address on the package, and it offers an easy starting point for the case.

The moral is to ensure that product packaging does not become implicated in safety violations or hazardous situations. This means that hazards analysis and other such assessments should be conducted for packaging as were conducted for the products themselves. These evaluations should identify all adverse consequences of normal and abnormal use, including abuse and foreseeable misuse for the entire lifetime of the packaging. The same requirements involving warnings of hazards that apply to products also apply to packaging. Foreseeable hazards associated with its handling or disposal should be explicitly described and prominently displayed, including corrective measures.

Plant Operations

Protecting the plant

A major resource. The function of an industrial manufacturing or process plant is to convert raw materials into refined forms and to assemble parts and components into useful and marketable products, structures, equipment, or entire engineered systems. As noted in the preceding section, the plant is responsible for more than the final product, as everything related to that product, including its interface with the user and even nonusers and bystanders, lies within the scope of the manufacturer's responsibility and duty.

Contrary to the traditional role of manufacturer in a previous era, responsibility for the product does not cease when it leaves the shipping platform. In many respects, the manufacturer's responsibility for the product becomes even more intense and of greater consequence after the product leaves the plant. However, the manufacturer's control over it diminishes significantly upon leaving its hands.

Of course, there are product recalls and some very limited contact with customers and users through service and warranties, but once the product has been released to the market, the product is on its own and the manufacturer's influence over its destiny is essentially over. It is a case of increased responsibility at a time when control is diminished.

What this means is that the product's life and its future, as well as the reputation of the manufacturer, are determined strictly while the product is within the manufacturer's control and still within its plant.

Since so much is riding upon what is done while the product is within the presence and control of the manufacturer, the plant, its operations, personnel, and facilities require close attention. These factors can have a major influence upon whether the product succeeds or fails, and there are many forms of failure as we have discussed.

The plant is the primary tool or resource of the business for converting incoming materials into valuable and marketable products. It is the instrumentality for financial income for the business. The plant will provide a return on the business investment if it is operated efficiently, if its expenses are minimized and controlled and if the plant can be trusted to perform as intended.

A plant failure or loss of a manufacturing component that is essential to its operation can interrupt production. This, in turn, causes product shipment delays and can adversely affect the business and its reputation. When the plant is down—and it need not be the *entire* plant, even a small segment of it can interrupt production—fixed costs continue and it becomes critical to restore its operation promptly. The continued health of the company is, therefore, directly tied to the continued health and vitality of the plant.

Plant availability. Discussion in Chap. 3 covered life-extension and life-cycle management issues related to major plant components. The goal was to maintain the efficient and effective operation of these components throughout their useful life, to avoid premature failures or forced retirement, and to ensure continued on-line availability.

Plant availability frequently does not receive the attention it deserves because it is less tangible and dramatic an issue than a forced shutdown due to a major failure. A manufacturing plant's availability can be seriously affected by a faulty component here and a balky component there, a few hours lost today and a few more tomorrow; all due to so-called minor occurrences.

It has been said by an executive of the Nuclear Regulatory Commission (NRC) that if U.S. nuclear plant availability could be improved by 10 percentage points, it would be the equivalent of constructing ten new nuclear plants.[80] No doubt the magnitude of improvement for manufacturing plants would be as dramatic. U.S. industry has only recently been awakening to the need to increase plant availability and to apply life-extension techniques that electric power utilities have been pursuing for some time.

These techniques, described earlier, do not constitute some new formula or miracle cure, but an organized and integrated program of consistently applying state-of-the-art knowledge to a critical situation. It is what is done on a day-to-day basis that can make all the difference between a mediocre plant and an efficient plant. The efficient plant has identified and characterized its susceptibilities and weaknesses; it has systematically analyzed them and monitors them. Through its detailed and readily retrievable and accessible information base, it knows how long critical components last and it schedules inspections and maintenance accordingly. In this way, it minimizes unscheduled shutdowns and production stoppages and avoids major failures.

The mediocre plant, on the other hand, is preoccupied with missed delivery schedules, panics over frequent equipment stoppages, operating budget overruns, poor employee morale, a high accident rate, and labor unrest. It operates on a "brushfire" mentality; that is, there is no time for preventive action amidst the problems. Its managers are "firefighters" and, instead of managing, run from blaze to blaze to stomp them out. Their regimen is little more than a string of production hangups, equipment malfunctions, switched priorities, and on-the-spot fixes and patches and reallocations of the labor force. Such an adverse situation feeds upon itself and can only worsen. It amounts to little more than a failure waiting for the "right" combination of circumstances to occur.

Performance. The approach to preserving and protecting one of the company's primary resources, the manufacturing or processing facility itself, can strongly influence its business performance. It is evident in its profitability, efficiency, and employee morale and its safety record and safety of its products. It is also evident in its excellent liability record. No manufacturer today is immune from liability claims, but those that operate responsibly with regard for safety of its products, its facility, and employees will have a strong basis for defending itself against those claims and for negotiating early and favorable settlements.

For purposes of evaluating integrity and reliability, the plant may be considered to be a large "product" having certain damage susceptibilities, hazards, defects, and weaknesses that can lead to failure. When viewed as a single entity, it is sometimes easier to analyze and evaluate and to identify critical flaws. The plant has an established and identifiable purpose and its effectiveness in fulfilling that purpose can be degraded by many things that can be identified or characterized. Its effectiveness may be determined through assessing its capability for converting incoming materials and labor into marketable and profitable products.[81] The analytical methods described earlier are readily adaptable to the task and can establish items and their priorities for management attention.

Since our emphasis is upon avoiding mechanical failures, those that concern the plant itself can affect (a) the ability of that resource to perform its intended function, (b) the optimal utilization of capital, and (c) the safety of its operation and its products. For the plant to satisfy its intended purpose, it must have maximum achievable availability, meaning a minimum number of unscheduled shutdowns. Optimal return on investment in the plant occurs when its effective operation can extend throughout the full term of its useful life. Safe operation and products ensure it will retain a larger portion of its income and that the expense of defending lawsuits, escalating insurance premiums, product recalls, and bureaucratic hassles will be minimal.

Plant effectiveness

Criticality identifications. Our interest here is in avoiding unscheduled downtime and the system faults and malfunctions that diminish plant availability and productivity. Analytical tools described earlier are indispensable in identifying critical faults and their causal elements so that they may be eliminated, controlled, or monitored to avoid the fault or failure. Before these techniques can be applied, however, there must be a method for determining what candidates require the most urgent attention. It is possible to conduct hazards and criticality analyses, or use other techniques, for the entire plant, then progressively work down into subsystems and components. And this should be done to ensure that all potential causes of failures have been considered.

Identification of production-critical equipment, however, requires a somewhat different approach. We are concerned here with ensuring maximum system availability. Failures and malfunctions certainly affect this, but so do other situations. Supplementing hazards and criticality analyses, FMEA, and FTA with process and schedule-oriented analyses can be helpful in identifying plant equipment and product subsystems that are vital to uninterrupted production.

Critical-path methodology (CPM), program evaluation and review techniques (PERT), and their variations have been used for some years in project planning and scheduling activities and in identifying sources of potential delays.[82] While their methodology has been applied most often for construction projects, they are adaptable to manufacturing and process operations. When coordinated with process or manufacturing flowcharts or networks, their time-oriented depiction of operations and events and scheduling constraints can provide valuable insights as to critical plant equipment, operations, and facilities.

If the CPM or PERT network is an accurate representation of the manufacturing system, it is possible to identify time-critical items or operations and associated equipment requirements. This information would assist in determining what is essential to uninterrupted production. Fault analyses of these would then identify components requiring primary attention. This procedure would help the plant operator to isolate from the multitude of possibilities the components most critical to the production operation and offer guidance in minimizing failures that would have the most detrimental effects.

Of course, many modern manufacturing facilities utilize sophisticated computerized process monitoring and control systems that accomplish these, and more, tasks. However, many smaller firms tend to regard such systems out of reach, as they may well be. Nevertheless, implementation of basic techniques need not be costly and yet can perform many of the same basic functions as sophisticated turn-key systems. In other words, the benefits of most of these techniques are available to all who would take the time to learn of them and apply them.

Improved perspective. A key to more effective plant operation is an improved understanding of process steps, material flow, scheduling, queuing, consequences of malfunctions and defective components, and many other factors. Existing operations often evolved from previous product lines and tradition without much planning, and those working with the system may not be aware of their deficiencies and inefficiencies. Process and material flowcharts, CPM networks, and such are system models, and these can be helpful tools in developing a perspective that will offer managers and operators the opportunity to identify aspects that need to be changed.

Models often provide a clearer understanding of the operation or process than studies of actual installations. If the model faithfully represents the real system, changes, modifications, and various tradeoff options can be readily evaluated at low cost without jeopardizing the actual system or compromising system availability.

Models can be of several types. They may be three-dimensional scale representations of the actual system or component, either smaller or larger, depending upon the item and application. They are useful in failure analysis, incident reconstruction, and assessing operational safety, as well as in process or product manufacturing layout evaluations for ensuring efficient material flow and other characteristics. Analog models substitute one property or set of properties for another, as in flowcharts, graphs, and other pictorial representations. Analytical procedures described earlier make extensive use of symbolic models of relationships among system components and other factors.[83,84]

Monitoring and control. Process and manufacturing equipment that is prone to malfunction and potentially responsible for product losses and delays should be studied using an appropriate method. Depending upon the kind of equipment and nature of the problem, it may be a failure analysis type of study, FMEA, or FTA. This is a preferred approach to repetitive repairs and replacements that do not address the root of the problem.

If it turns out that demands being placed upon the unit are such that certain components cannot be expected to last for an acceptable interval and if a long-term solution is not available, relief may still be possible through monitoring. Depending upon the component and failure mode, inexpensive sensors may be placed upon the component or its output to signal approaching failure. Then the item can be replaced at a convenient time when equipment can be shut down without disrupting production. The given situation may not be so simple, and the failure mode may not be readily discernible at a prefailure stage; however, there usually is some quality or performance characteristic that can be monitored to provide early warning of impending problems. If not, a redundancy arrangement should be considered.

Conditions that can severely degrade a process should, similarly, be monitored. Real-time process control is an increasingly popular theme, and a variety of systems for virtually any production or process situation have been or are being developed.

Manufacturing process control has been practiced for years; however, it is usually done at some interval after deviations have been observed. In a typical situation, final product quality inspection will note a need for correction of some aspect. This information is communicated back to the respective manufacturing stations where corrections are made and their outcome evaluated until desired results are obtained. Meanwhile, material processed during out-of-control intervals may require rework or be scrapped.

This is process control, but it is not efficient. If, however, the inspection step could be moved closer to the actual production operation and deviations observed and corrected as they occur, the time interval between occurrence and deviation would be shortened. Then the number of products produced under faulty conditions would be decreased, and control efficiency would increase.

This goal may be achieved in various ways. One is for sensors to monitor the operation or processed part for compliance with some desired characteristic, dimension, or quality standard. Deviations observed could be signaled to an operator who makes the correction. The more sensitive the monitoring method is to the deviations, the less correction would be needed. A more sophisticated scheme might use a real-time feedback and control arrangement whereby observed deviations and their magnitude are sensed and communicated back to process control and translated in terms of machine settings adjustments required to correct the deviation, for example.

Such closed-loop control would combine monitoring and control functions that would fine-tune manufacturing operations and correct minor deviations as they occur. The system would require capability to sense, analyze, make decisions, and implement corrective action. If it could be worked out and be reliable, such an arrangement would have obvious advantages, and this is the goal of much of this research.

Depending upon the operation being controlled, the control system might be extremely complex. Accordingly, the usual evolution of such systems occurs a step at a time. Sensors are the first element required and these are being implemented.[85–89] Analysis of the indications, the decision-making process, and implementation of corrections are performed manually in most situations today. The advantages over relying upon final product inspections are obvious, particularly for complicated processes and environments or prolonged operations requiring constant attention and adjustment. As progress is made in applying computerized systems to analysis, decision making, and corrective implementations in on-line manufacturing control, the number of rejects, defective products, and hazardous processes will no doubt diminish significantly.[90–93]

Control reliability. Regardless of the appeal and promise of manufacturing control systems, their reliability must be thoroughly evaluated and confirmed over a wide range of operating conditions before entrusting them to actual use. Even then, safeguards must be built in that can sense controller malfunction and either correct the problem, return the system to a safe standby condition and alert operating personnel as to the system status, or revert to manual control or some kind of fail-safe condition.

When contemplating use of on-line controllers, it is important not to exchange one failure-susceptible component or system for another. Manufacturing system reliability or availability before and after implementing process control must be evaluated, and the evaluation should include realistic failure rate information for the controlling elements. This applies not only to complete control systems, but for all elements, including sensors. However, these are usually most vulnerable to damage and require the most attention. If results indicate that controller faults, malfunctions, and failures will degrade the process and its availability, and yet on-line control is essential, fault tolerant control designs and procedures may be required.[94]

Fault tolerant control systems rely upon some form of redundancy to maintain system availability. The type and degree of redundancy depend upon the controller component and the kind of fault. Faults in controller components, like other products or components, may be traceable to such causes as inadequate or defective design or specification errors or omissions, to manufacture, to component failure, or to unanticipated environmental extremes. As discussed for other systems, avoidance of controller faults or failures can be approached through implementing the various analytical procedures such as hazards and criticality analyses, FMEA, FTA, design reviews, NDE, and other methods.

Fault tolerance means that the system has provisions for avoiding major perturbations that would diminish control effectiveness and degrade the process being controlled. It is a fail-safe approach that allows system faults to exist without degrading performance; it does not attempt to eliminate them. As noted earlier, redundancy in some form is usually applied to accomplish this. However, before redundancy can be built into the system, there must be recognition of the fault, its location must be determined, it must be isolated and contained, and system recovery effected. Redundancy ensures that when a control system failure occurs, there are sufficient provisions within the system for continuing satisfactory operation.

Provisions of controller redundancy may take several approaches: software, time, or hardware. It does not always involve component duplication, but frequently does. Software redundancy involves supplemental systems to detect and characterize control system faults. Time redundancy employs reexecution procedures in attempting to correct the apparent error. Hardware redundancy is replication of fault-susceptible components and parts having the highest failure rates.

Effectiveness of the fault tolerant design should be evaluated using appropriate system reliability modeling techniques incorporating probabilities (e.g., Markov modeling).[95,96] If these reveal inadequate or unsatisfactory behavior, further control system modifications using more reliable components or more fault tolerant designs are required.

Plant maintenance. It is well known that time diminishes the reliability of mechanical and other engineered components. Time spent in actual operation is sometimes a useful gage for evaluating integrity or reliability, but time in existence or age also must be considered. This is regardless of time in actual use. Maintenance is periodic inspection and care given to equipment or components to ensure their continual availability and operation at the intended level of performance.

The operational integrity of a plant is directly related to the caliber and consistency of its maintenance program. All equipment, monitors, control instrumentation, and peripherals require maintenance and their reliability depends upon it.

Maintenance may be corrective, preventive, or predictive in nature. A corrective maintenance approach attends to components when they need attention. It is really maintenance by default—the "if it's not broken, don't fix it" philosophy. In a strict sense, this is not really *maintenance.* It ignores these issues and only attends to equipment when it wears out, fails, or no longer adequately serves its intended purpose. In some respects, this method can have advantages. Periodic disassembly for preventive maintenance can incur risks of errors, misplaced and incorrectly replaced parts, and other problems. And it can be economical for some components to be discarded and replaced with new when they fail. This is the trend for many household and consumer products and others.

However, this approach to maintenance for most industrial components can cause major problems when failures and malfunctions disrupt operations, curtail or interrupt production, or cause injuries and other damage. In these situations, costs of premature replacement of the item, or a preventive maintenance approach can be insignificant in comparison to the consequences of its failure. Therefore, this approach is used only infrequently or for components designed to be "maintenance-free" and simply discarded when they malfunction or have reached the end of their useful life.

Preventive maintenance is the traditional standard approach. Periodically, the unit is serviced according to the manufacturer's or other recommendations, even if it is operating satisfactorily and may have had no previous record of difficulty. It is time consuming, can be expensive, and can introduce problems that had not existed before. These are usually the reasons or excuses for choosing the corrective maintenance approach. Preventive maintenance is based upon the "an ounce of prevention is worth a pound of cure" adage. But for expensive equipment that has been trouble-free and a plant with thousands of items to handle, its significant costs can be difficult to swallow.

The predictive maintenance approach attempts to capture the best of both worlds in combining, to some degree, corrective and preventive practices. It is corrective maintenance in the sense that the component is not disassembled and checked and repacked with lubricant and its seals replaced, for example, unless it shows symptoms of impending problems. This scheme requires adequate and effective diagnostic monitoring to detect such conditions, plus a means for relaying this information to someone who can attend to it or schedule corrective action for the next available opportunity.

This method can save the costs of periodic disassembly or whatever action is taken regardless of operating equipment status or condition. But these must be weighed against monitoring and information feedback costs and risks associated with monitor malfunctioning. Also, the monitoring technique used must be capable of diagnosing the correct and appropriate equipment fault.

Whichever approach is chosen—preventive or predictive—computer-assisted maintenance scheduling and diagnostic information storage techniques can offer benefits in keeping ahead of the servicing needs of a large plant. Even small plants can benefit from these systems, which can ensure that all equipment is accounted for and is scheduled for attention at appropriate intervals.

The software system may be no more than a scheduling tool or it can be much more. It can also keep track of maintenance parts inventory and assist in purchasing optimal units for price breaks and in maintaining realistic quantities of spare parts and supplies. Predictive maintenance software can store and analyze diagnostic data such as operating temperatures, vibration signatures, and spectrographic lubricant analysis results for disclosing operational trends that can reveal needs for correction, repair, or replacement. Progress in developing more discriminating and robust sensors and NDE refinements in such techniques as infrared thermography, ultrasonic testing, and other methods should further improve plant availability in a cost-effective and timely manner.[97,98]

Optimal utilization of capital

Engineering design of industrial equipment is based upon conservative practices to avoid failures during the projected operating lifetimes of the equipment. This is necessary because of uncertainties of operating conditions, the inability to predict material response to those conditions accurately, and the allowance for inherent flaws in materials and components and for undetected fabrication errors and defects.

Conservative design practices incorporate factors of safety, calculations based upon minimal mechanical properties and worst-case environmental and use scenarios, and conservatism in operating specifications, maintenance recommendations, and service and inspection intervals. The trend toward increasing manufacturer liability during the last decade or so has driven equipment manufacturers and others to even greater conservatism. The result of all of this is that the usable lifetime of major pieces of equipment may be double or triple that of its original "design life."

In view of escalating costs of new plant construction, associated regulatory and environmental restrictions, and other negative factors, there are compelling incentives for operating plant equipment beyond its originally intended design life, as noted in Chap. 3. Such decisions are not easy to make. There are risks involved and there is the potential difficulty of justifying extended operation of equipment and components that have exceeded their so-called design life. If failures occur and are traced to deliberate decisions to operate the equipment beyond its design life, it could be damaging from a liability viewpoint when continued operation was shown to be economically motivated. This option can pose a problem for insurers, as well. Consequently, decisions for extended operation must be based upon sound engineering principles, defensible techniques, and well-documented analyses and implementation practices.

Even though the original designs may have been conservative and state-of-the-art safe practices applied at the time of their construction, intervening factors may have had adverse effects. For one thing, design practices at the time of equipment construction may, in the light of present knowledge, not have been as conservative in some respects as originally believed. Modifications made since the initial installation may have defeated builtin safety factors, increased operating stresses, intensified stress concentrations, increased operating cycles and temperatures, aggravated environmental factors, and may have introduced many other such conditions that could affect equipment life. These possibilities tend to counterbalance and can outweigh the original conservative design approach.

For these reasons, any consideration of extending the operating life of plant equipment must involve assembling all available information on the operating, repair, and maintenance history of the equipment. This should be supplemented with direct assessments of its condition. (See Chap. 3 for more details.) It is important that these assessments are carried out with full understanding of susceptibility of the equipment to various damage mechanisms and failure modes, and these should be identified, classified, and rated with respect to their criticality and likelihood using appropriate analytical techniques.

Condition of the equipment and magnitude of accumulated damage by the most likely and other possible damage mechanisms should be determined and assessed in terms of remaining life. When this has been done, continued operation may be justified without restriction, or the decision may be made to decrease operating stress, cyclic loading, temperature, or some other condition as indicated by analytical test results. It is possible that critical components should be retired and replaced with state-of-the-art units.

During continued operation, critical regions should be monitored for progressive damage and appropriately short inspection intervals established that will permit potentially detrimental conditions to be detected before the onset of failure. A good deal of engineering judgment is involved in all of these decisions, but judgment that is firmly based upon solid test data on material condition and performance status of the subject equipment.

Plant safety

Failures of plant equipment can be damaging in the risks of injury and harm to operating personnel as well as to those who are not directly associated with the operation but who happen to be located in its vicinity. In recent years, environmental effects of industrial mishaps have become increasingly significant. The liability of plant operators for injuries resulting from accidents, failures, and mishaps is evident from news accounts of incidents, some having extensive and severe consequences. Damage to equipment and adjacent facilities and operations can also result in diminished production output or complete shutdown, with accompanying loss of revenue and other damage to the business and its reputation. Therefore, there are major incentives for avoiding plant failures.

Whether or not there is interest at the moment in extending the operating life of major equipment, safety analysis of the plant is not optional, particularly in today's safety conscious and litigious society. In fact, safety issues should be part of all hazards assessment and other analytical studies conducted in evaluating plant condition and availability. Increased regulations aimed at improved employee safety make many of these practices mandatory. As discussed in the opening chapter, people-related elements are a major contributor in most failure incidents, whether they occur in the plant or elsewhere, and human factors rank high on the list of things requiring attention.

Plant safety includes consideration of many of the factors involved in assessing product safety, but some aspects require greater emphasis. Manufacturing and processing plants can be extensive operations involving massive equipment and facilities, large numbers of people, sizable quantities of raw material, and large amounts of energy. When these get out of control, damage can be extensive. Consequently, hazards assessment, safety practices, and damage minimization procedures for plants require additional emphasis in some aspects.

Because of the dominant effects of human behavior and response, these factors must receive priority attention. The potentially serious consequences of plant accidents and failures require study to characterize these consequences and to develop techniques for mitigating them. This would include dispersion analyses for accidental releases of toxic fumes, dusts, smoke, and other gases and substances, and studies of the probable extent of damage from explosions and fires, including their secondary effects on the immediate location and surroundings.[99,100] Depending upon the plant, its size, and nature of its operations, organization of an emergency response or incident control activity might be appropriate.

Some manufacturing and related operations are not confined to plant boundaries and can extend for large distances as raw materials, reagents, or products are transferred by truck, ship or barge, and pipeline. This is often done through heavily populated areas and environmentally sensitive locations. This extension of risks must be considered and assessed and adequate handling precautions developed and implemented with procedures for minimizing damages from accidental spills and transportation equipment failures.

Such issues must include disposal of by-products, wastes, spent reactants, scrap, and all other substances and materials associated with the plant and its operations. It is well to recognize that liability may not transfer with ownership or title, and manufacturers responsible for the existence of a material may be held liable for harm and damage it causes, long after the material has been transferred to other locations or individuals.

Because of the very significant risks associated with plant operation and the heavy responsibility imposed upon plant operators, the design of control equipment, user-interface aspects, and operator training are increasingly important factors in avoiding problems and incidents. The legal responsibility for operation and control of a major facility can often exceed that of commercial aircraft pilots, and yet the formalized training requirements for the pilot are often much greater than for plant operators or process control personnel.

Control room operators are usually trained by experienced people who may be retiring or reassigned and the information is passed along informally during an indoctrination period. Operators are trained to monitor various indicators and to appropriately respond to these indications. However, their understanding of the complex process or operation may not be adequate to handle abnormal situations or unpredictable response due, possibly, to process faults, monitor malfunctions, or component failure. As noted earlier, the Three Mile Island nuclear plant incident stemmed from an incomplete understanding of the operation, incorrect operator assessment, and confused response to control indications. Increasing use of plant control panel simulators, along the lines of aircraft control simulators, should assist in decreasing the frequency of such incidents.[101]

Life-extension activities sometimes involve system upgrades and installation of new components that require different forms of response, indicator reactions, and control procedures. Also, older generations of control or monitoring technology can suffer from degraded response and other reliability problems. These situations require reassessment of the capabilities of the control system, its logic, display format, and operator interface.[102]

The Right Corporate Climate

Risk management

It has been said that technology has never injured anyone or damaged anything, but mismanagement of technology has been at the root of all failures, accidents, and mishaps. As emphasized earlier, failures stem from human deficiencies. Analyses of the causes of failures, including the major catastrophes of recent years such as Chernobyl, Three Mile Island, Bhopal, and the *Challenger* space shuttle disaster, point to management errors, not failure of technology. While we may be aware of this, there is a tendency to ignore the obvious and to be reluctant to admit to human frailties; therefore, it is easy to lose sight of the dominant management role in failure avoidance. To get it done, it takes more than a group of designers, production engineers, and all the other employees of an organization with awareness and understanding of how to make machines and products safe, and even dedication to safe practices.

As has been discussed, failures of a company's products and equipment within its manufacturing facilities can put it out of business. Therefore, these issues deserve, and demand, the primary attention of top management. All businesses have risks, and these must be carefully and thoroughly assessed, identified, and understood, and decisions made for their disposition. Ignoring them is not a viable option. There are proven techniques for assessing risks and some of them have been described earlier. Assessment and characterization of risks inherent in operating the business are indispensable to intelligent and informed operation.

In general, there are four available options for handling business risks. They are (a) elimination, (b) retention, (c) transfer, and (d) reduction. Some options are more appropriate to some risks than others, and some options may be undesirable.

Risk elimination. The option of risk elimination usually means discontinuing the business activity or at least the activity that incurs the risk. If there is a high risk of product liability in manufacturing a product that has questionable utility and a propensity for failure with associated damage or potential for personal injury or other products having the same or similar functions are safer to use, it might be prudent to eliminate the risks. This might be done through discontinuing manufacture of the product or, preferably, by not even beginning manufacturing it.

This is often the basis for the inherently safe approach—eliminate the element or elements that can cause failure. It is also the motivating force behind the desire for keeping things as simple as possible. What is not there cannot fail, fracture, fatigue, leak, explode, burn, decompose, short out, deteriorate, corrode, or damage anything. Similarly, people that are not there, in the way, or in the area cannot be injured or file liability claims. Risk elimination is a most effective option, and it is not considered often enough. But there are many situations where it is not possible. Businesses must be conducted, products must be manufactured and marketed, and all risks cannot be eliminated.

Risk retention. It may be decided that some business risks must be assumed, retained, or simply accepted. This may be termed *self-insurance* or the willingness to accept the risk and absorb any related losses. There are various reasons for doing so, and the decision may be sound or unsound. It would be unsound if the risk was incorrectly assessed or the decision was based upon erroneous assumptions or miscalculations.

Circumstances change, along with society's values, the legal and regulatory climate, the market, and other factors. Therefore, risk assessment cannot be a one-time exercise, but an ongoing activity. Continual monitoring of business, marketing, and economic conditions is required. Reevaluations and reassessments of the firm's affairs and its activities must be considered in the light of current and projected developments. What may be a reasonable risk to assume today may become an unacceptable one tomorrow.

Risk transfer. Risk transfer may be accomplished through insurance or contractual agreement. We have already observed the limited value of disclaimers in a manufacturer's attempt to avoid the risk of liability in the marketing of consumer products. It was noted that a manufacturer's obligation to make safe products could not be contracted away, particularly not to unwary customers whom the courts view as helpless victims of "big business." However, in commercial transactions, there is more leeway for risk transfer.

Transfer of liability risks has been a key element in the products-liability "crisis" of the past decade or so. In fact, leading court decisions that were stepping stones in the progressive evolution of products-liability law in general, and in the development of strict liability principles in particular, reconciled the decision to impose these obligations upon manufacturers in citing their ability to transfer these risks:

> [T]he burden of losses consequent upon use of defective articles is borne by those who are in a position to either control the danger *or make an equitable distribution of the losses when they do occur.* (*Henningsen v. Bloomfield Motors, Inc.*[103]) (emphasis added)
>
> The cost of an injury and the loss of time or health may be an overwhelming misfortune to the person injured, and a needless one, *for the risk of injury can be insured by the manufacturer and distributed among the public as a cost of doing business.* (*Escola v. Coca Cola Bottling Co. of Fresno*[104]) (emphasis added)

As history has shown, the presumed unlimited availability of insurance coverage for manufacturers to shift products-liability losses was overstated, at best. During the 1970s in the face of increasing liability claims and escalating damage awards, many manufacturers found themselves underinsured, unable to afford the sharply increased premiums, or uninsurable. A study conducted in 1976 found that during the period of 1971 to 1976 the average damage award per firm surveyed increased in value from $434,000 to over $3.5 million.[105] The same study showed that from 1965 to 1974 manufacturers' liability claims increased 1300% while the Consumer Price Index had risen only 56%. During this period, machine tool builders experienced an average increase in products-liability insurance premiums from $10,000 in 1970 to $140,000 in 1979. There were many reasons for this situation, not the least of which were the multiplicity of state laws that varied from place to place, and uncertainties in how courts would react in these jurisdictions, making it difficult for insurers to assess their exposure. Increased damage awards and the unsettled question of coverage for punitive damages did not help.

Accordingly, the availability and affordability of liability insurance, or the ability of manufacturers to transfer these risks via the purchase of insurance coverage, were responsible for much apprehension and expense for product manufacturers. They contributed to the woes of many and prompted the demise of a number of product lines—the small aircraft (general aviation) business, to name one.

The lesson in all of this for manufacturers a decade or more later is that the option of insurance as a means for transferring risks should not be considered a primary approach to the problem of liability for harm and damage caused by accidents, mishaps, and failures of manufactured products. Insurance is necessary protection and businesses should have adequate coverage; however, it can be an elusive protector.

The manufacturer must strive for an optimal balance among the several approaches for controlling risk. The right combination will differ with the nature of the business and its operations, products, liability exposure, and industry factors.

Risk reduction. The approach of risk reduction is the subject of this book, and it can represent the most effective method, short of eliminating the product line or business altogether. This is because, in reducing the risk of failures in the plant and in its products, a manufacturer accomplishes other very significant things, as well. Reduced risks mean fewer claims, lower insurance costs, less bureaucratic interference, a more efficient operation, better employee morale, higher profits, better public relations, greater investor confidence, and other positive effects and intangibles such as peace of mind.

It is not obtained without cost, however. Manufacturers must assess their own situations and determine the nature and magnitude of the risks and the right combination of approaches for mitigating them, yielding the most cost-effective balance.

The failure avoidance strategies described in these pages are appropriate for, and have emphasized, high-risk activities, as this is where the need is greatest. High-risk products and plant operations require full implementation of all means for controlling these risks and minimizing damaging failures. On the other hand, low-risk operations and products will not require the same extent of failure avoidance activity. Although no analytical method is a substitute for sound judgment, the various techniques described can be useful aids in exercising that judgment. They are merely tools to make the tasks easier and be accomplished more effectively, and we can all benefit from their use.

Management's role

Every company will differ from the rest, along with its objectives, policies, risks, operating style, and optimal approach to managing plant and product safety, but there will be common requirements. For example, no matter how large or small the plant and its risks, failure avoidance issues must be considered on a company-wide basis. Every operation, employee, and product made and sold must be involved. The extent of involvement and approaches used, and so on, will depend upon many factors, but, whatever is done along these lines—if there is to be any significant effect—must include everyone.

It is not so much a policy which can be reduced to printed words on paper as it is a climate. Activity or operating style, or whatever it may be called, that is so pervasive and significant to a company's vitality and future must originate with and receive its continued support and

- Designate organizational authority for failure avoidance in its products and plants, and define responsibilities for minimizing associated risks
- Prepare, disseminate, and implement an organizational failure avoidance policy
- Provide resources for implementing this policy
- Provide and implement appropriate design, material, and component selection and procurement standards
- Designate responsibility to ensure compliance with regulatory and legal standards
- Provide educational programs to develop understanding consistent with failure avoidance responsibility
- Audit the failure avoidance program

Figure 4.2 Management implementation steps in organizing a failure avoidance program.

sustenance from top management. This is absolutely essential as anything less will fail.

Any attempt to define general management tasks required for implementing a corporate-wide failure avoidance or product and plant safety program will be lacking, as it must deal in nonspecifics and cover a broad spectrum of types of operations and situations. Nevertheless, various attempts have been made to do this and one version is shown in Fig. 4.2.

The listed order is generally chronological but not rigidly so. Some are one-time tasks while others are ongoing activities. In implementing such a program or plan, there is an obvious need for good communication throughout the organization, as many of these elements involve virtually every operation. Implementation will not occur instantly, by edict, as it must be a gradual educational process. How it is phased in and its schedule will depend upon the company and the nature of its business and many variables.

It is best to have a formal plan and schedule that is formulated and endorsed by line managers. This will not only ensure its implementation but will minimize disruptions. The plan should not be hastily conceived or developed and must accommodate every aspect of the operation. It should receive, as a bare minimum, the same analysis, scrutiny, and consideration as a major new product or product line. And many of the techniques used for such evaluations, including decision analyses and design reviews, may be applied.

Audits

Management audits of the activity are essential to determine if the program is taking root and whether it is on track and progressing according to plan. Everyone must understand that the audit is not an investigation but a planned review.

Elements of the audit should include (a) identification of factors to be reviewed, (b) an audit plan for each function, (c) an audit schedule, (d) arrangement or mechanism for reporting results and outcome, and (e) consultation to discuss recommendations and corrective action. The audit process should not be deferred until the plan is complete and in place, as the overall plan, schedule, and implementation must be reviewed frequently to ensure that it is progressing as intended and to resolve conflicts and difficulties.

References

1. Patricia L. Ricci, "Standards Sources and Resources," *ASTM Standardization News,* June 1990, pp. 54–59.
2. *Final Report of the National Commission On Product Safety,* Superintendent of Documents, U.S. Government Printing Office, 1970.
3. *American Law of Products Liability 3d,* The Lawyers Co-Operative Publishing Co., Rochester, NY, 1987. Set of 15 binders, supplemented and updated quarterly, includes monthly newsletter *Products Liability Advisory.*
4. *Product Safety and Liability Reporter,* The Bureau of National Affairs, 1231 25th St., N.W., Washington, DC 20037. Annual subscription fee provides binders and weekly reports.
5. *The Federal Register,* U.S. Government Printing Office, Washington, DC 20402.
6. *Index to Government Regulations,* The Bureau of National Affairs, 1231 25th Street, N.W., Washington, DC 20037.
7. *Environmental Reporter,* The Bureau of National Affairs, 1231 25th Street, N.W., Washington, DC 20037. Annual subscription fee provides binders and weekly reports.
8. *Products Liability Reporter,* Commerce Clearing House, Inc., 425 13th Street, N.W., Washington, DC 20004.
9. *Loss Prevention and Control,* The Bureau of National Affairs, 1231 25th Street, N.W., Washington, DC 20037. Annual subscription fee provides binders and weekly reports.
10. *Patty's Industrial Hygiene and Toxicology* (3 volumes), Wiley, New York, NY, 1978–1982.
11. *NIOSH Registry of Toxic Effects of Chemical Substances (RTECS),* Superintendent of Documents, U.S. Government Printing Office, Washington, DC 20402.
12. *Phillips v Kimwood Machine Co.,* 525 P.2d 1033 (Oregon, 1974).
13. *Barker v Lull Engineering Co.,* 20 Cal.3d 413, 143 Cal. Rptr. 225, 573 P.2d 443 (1978).
14. S.44, 98th Congress, 1st Session, Sec. 2(6) (1983).
15. Edward M. Swartz, *Proof of Product Defect,* Lawyers Co-Operative, Rochester, NY, 1985, Sec. 2:1.
16. Reference 13.
17. *Voss v Black & Decker Manufacturing Co.,* 59 N.Y.2d 102, 463 N.Y.S.2d 3998, 450 N.E.2d 204 (1983).
18. *Restatement (Second) of the Law: Torts,* Sec. 402A, Comment i, American Law Institute Publishers, St. Paul, MN, 1965.
19. Ibid., Comment k.
20. Reference 14.
21. Reference 2.

22. Charles E. Witherell, *How to Avoid Products Liability Lawsuits and Damages—Practical Guidelines for Engineers and Manufacturers,* Noyes, Park Ridge, NJ, 1985.
23. Irwin Gray, *Product Liability—A Management Response,* American Management Association, New York, NY, 1975.
24. Alvin S. Weinstein, Aaron D. Twerski, Henry R. Piehler, and William A. Donaher, *Products Liability and the Reasonably Safe Product—A Guide for Management, Design, and Marketing,* Wiley, New York, NY, 1978.
25. Charles O. Smith, *Products Liability: Are You Vulnerable?,* Prentice-Hall, Englewood Cliffs, NJ, 1981.
26. James F. Thorpe and William H. Middledorf, *What Every Engineer Should Know About Product Liability,* Dekker, New York, NY, 1979.
27. Charles E. Witherell, "The Products Liability Threat," *Chemical Engineering,* Vol. 90, No. 2, January 24, 1983, pp. 72–87.
28. Edward M. Swartz, Ref. 15, p. 12 (Sec. 2:3).
29. Harris R. Greenberg and Joseph J. Cramer (eds.), *Risk Assessment and Risk Management for the Chemical Process Industry,* Van Nostrand Reinhold, New York, NY, 1991.
30. H. D. Voegtlen, "New Product Quality," *Quality Control Handbook* (J. M. Juran, Frank M. Gryna, and R. S. Bingham, Jr., eds.), 3d ed., McGraw-Hill, New York, NY, 1979, pp. 8-53–8-56.
31. John Kolb and Steven S. Ross, *Product Safety and Liability—A Desk Review,* McGraw-Hill, New York, NY, 1980, p. 116.
32. Willie Hammer, *Handbook of System and Product Safety,* Prentice-Hall, Englewood Cliffs, NJ, 1972, pp. 321–342.
33. William W. Doerr, "'What If' Analysis," in Ref. 29, pp. 75–89.
34. Willie Hammer, Ref. 32, pp. 156–159.
35. R. M. Sherrod and W. F. Early, "Hazard and Operability Studies," in Ref. 29, pp. 101–126.
36. Ernest J. Henley and Hiromitsu Kumamoto, *Reliability Engineering and Risk Assessment,* Prentice-Hall, Englewood Cliffs, NJ, 1981, pp. 8–43.
37. A. L. Dorris and J. L. Purswell, "Warnings and Human Behavior: Implications for the Design of Product Warnings," *Journal of Products Liability,* Vol. 1, No. 4, 1977, pp. 255–263.
38. K. Ross, "Legal and Practical Considerations for the Creation of Warning Labels and Instruction Books," *Journal of Products Liability,* Vol. 4, No. 1, 1981, pp. 29–45.
39. P. Sperber, "The Strategy of Product Labeling for Loss Prevention," *Journal of Products Liability,* Vol. 1, No. 3, 1977, pp. 171–182.
40. Edwin R. Boquist, "Tutorial on Formal Design Reviews," *Proceedings PLP-73,* Product Liability Prevention Conference, August, 1973, pp. 75–84.
41. Ibid., pp. 77–80.
42. J. C. Lewis, "Application of Fracture Mechanics to Spacecraft Design," *Fracture and Failure: Analyses, Mechanisms, and Applications* (Paul P. Tung, S. P. Agrawal, A. Kumar, and M. Katcher, eds.), American Society for Metals, Metals Park, OH, 1981, pp. 1–14.
43. *Damage Tolerant Design Handbook—A Compilation of Fracture and Crack Growth Data for High Strength Alloys,* Metals and Ceramics Information Center, Battelle-Columbus Laboratories, Columbus, OH, 1975.
44. G. C. Sih, *Handbook of Stress-Intensity Factors for Researchers and Engineers,* Institute of Fracture and Solid Mechanics, Lehigh University, Bethlehem, PA, 1973.
45. H. Toda, P. C. Paris, and G. R. Irwin (eds.), *Stress Analysis of Cracks Handbook,* Del Research, Hellertown, PA, 1973.
46. Ward D. Rummel, "Application of Nondestructive Evaluation in the Aerospace Industry—Reliability and Cost Impact," *Prevention of Structural Failures—The Role of NDT, Fracture Mechanics, and Failure Analysis,* American Society for Metals, Metals Park, OH, 1978, p. 319.

47. Royce G. Forman and Tianlai Hu, "Application of Fracture Mechanics on the Space Shuttle," *Damage Tolerance of Metallic Structures: Analysis and Methods and Applications, ASTM STP 842,* American Society for Testing and Materials, Philadelphia, PA, 1984, p. 112.
48. N. E. Frost, K. J. Marsh, and L. P. Pook, *Metal Fatigue,* Clarendon, Oxford, England, 1974, p. 204.
49. W. H. Lewis, W. H. Sproat, and M. W. Pless, "Quantitative Measurement of the Reliability of Nondestructive Inspection on Aircraft Structures," Ref. 46, pp. 164–183.
50. Alten F. Grandt, Jr., "Introduction to Damage Tolerance Methodology," Ref. 47, p. 16.
51. Stanley T. Rolfe and John M. Barsom, *Fracture and Fatigue Control in Structures—Applications of Fracture Mechanics,* Prentice-Hall, Englewood Cliffs, NJ, 1977, p. 261.
52. Alten F. Grandt, Jr., Ref. 47, pp. 3–24.
53. Stanley T. Rolfe and John M. Barsom, Ref. 51, pp. 260–264.
54. James L. Rudd, "Air Force Damage Tolerant Design Philosophy," Ref. 47, pp. 134–141.
55. Stanley T. Rolfe and John M. Barsom, Ref. 51, pp. 292–309.
56. Willie Hammer, Ref. 32, pp. 259–261.
57. Ibid., pp. 283–286.
58. Stanley T. Rolfe and John M. Barsom, Ref. 51, pp. 362–365.
59. Carl C. Osgood, *Fatigue Design,* Wiley-Interscience, New York, NY, 1970, pp. 23–25.
60. James L. Rudd, Ref. 54.
61. Ibid., p. 134.
62. Royce G. Forman and Tianlai Hu, Ref. 47, p. 108.
63. Stanley T. Rolfe and John M. Barsom, Ref. 51, pp. 431–440.
64. "Proposed Uniform State Product Liability Act," National Product Liability Council, text published in *Proceedings PLP-78,* Product Liability Prevention Conference, Philadelphia, PA, August, 1978, pp. 169–194.
65. Lee O'Connor, "Controlling the Turn of the Screw," *Mechanical Engineering,* September, 1991, pp. 52–57.
66. Michael Valenti, "Stemming the Flood of Counterfeit Fasteners," *Mechanical Engineering,* February, 1993, pp. 46–49.
67. Girard S. Haviland, "Designing With Threaded Fasteners," *Mechanical Engineering,* October, 1983, pp. 17–31.
68. Robert O. Parmley (ed.), *Standard Handbook of Fastening and Joining,* McGraw-Hill, New York, NY, 1977.
69. E. Paul DeGarmo, *Materials and Processes in Manufacturing,* 4th ed., Macmillan, New York, NY, 1974, p. 8.
70. Joseph Constance, "Can Durable Goods Be Designed for Disposability?" *Mechanical Engineering,* June, 1992, pp. 60–62.
71. *Uniform Commercial Code* (*UCC*), Uniform Laws Annotated (Master Edition), Secs. 2-312 to 2-318, 2-607, 2-718, 2-719, 2-725.
72. *Magnuson–Moss Warranty—Federal Trade Commission Improvement Act,* Public Law No. 93-637, 93rd Congress, S.356, 15 USC 2301-12, January 4, 1975; also, *Interpretation of Magnuson—Moss Warranty Act,* 16 CFR 700, 42 FR 36112 (July 13, 1977).
73. Ibid., p. 108.
74. Steven Ashley, "Handle With Care: Designing Damage-Proof Packaging for Products," *Mechanical Engineering,* October, 1992, pp. 66–70.
75. *Restatement* (*Second*) *of the Law: Torts,* Section 402A, Comment g, American Law Institute, St. Paul, MN, 1965.
76. H. L. Kaplan and R. E. Fawcett, "Components of Manufacturers' Products Liability Based Upon Defective Packaging: Foreseeability, Superseding Cause and Federal Preemption," *Journal of Products Liability,* Vol. 7, No. 2, 1984, pp. 119–141.

77. W. M. Carley, "Johnson and Johnson Is Hit With First Suit Following Deaths from Poisoned Tylenol," *The Wall Street Journal,* October 5, 1982, p. 24, col. 2.
78. H. M. Berg and R. A. Kosseff, "Should the Manufacturers Be Held Liable for the Tylenol Murders?" *For the Defense,* Vol. 24, December, 1982, p. 12.
79. *47 Federal Register* 50442 (1982), as amended, *48 Federal Register* 1706 (1983). 21 CFR, Secs. 211.132, 700.25, and 800.12 (1983).
80. David I. Lewin, "Washington Window," quoting Victor Stello, NRC Executive Director for Operations, *Mechanical Engineering,* January, 1988, p. 60.
81. Chan C. Park, "Counting the Costs: New Measures of Manufacturing Performance," *Mechanical Engineering,* January, 1987, pp. 66–71.
82. R. L. Martino, *Critical Path Networks,* McGraw-Hill, New York, NY, 1967.
83. Willie Hammer, Ref. 32, pp. 95–96.
84. Nicholas K. Mango and Stan Knutson, "A Computer-Aided Dynamics System Modeler," *Mechanical Engineering,* July 1985, pp. 52–63.
85. Howard D. Haynes, "Check Valves: Oak Ridge's New Diagnostics," *Mechanical Engineering,* May, 1991, pp. 64–69.
86. David Horn, "Smarter Sensors Respond to Factory Stimulii," *Mechanical Engineering,* September, 1989, pp. 64–67.
87. Michael Valenti, "Sensors Unite for Process Improvement," *Mechanical Engineering,* September, 1992, pp. 54–58.
88. Lee O'Connor, "Vortex Meters: High-Accuracy Flow Measurement," *Mechanical Engineering,* October, 1991, pp. 40–45.
89. Michael Valenti, "Infrared Sensors—Hands-Off Temperature Measurement," *Mechanical Engineering,* October, 1991, pp. 40–45.
90. Sheree Wen, "An Intelligent Path To Quality—Process Monitoring and Control," *Journal of Metals,* January, 1991, pp. 10–12.
91. Bevan P. F. Wu, "The Application of Expert Systems To Process Control," *Journal of Metals,* January, 1991, pp. 13–16.
92. Kellyn S. Betts, "Process Control Takes Command," *Mechanical Engineering,* July, 1990, pp. 64–68.
93. David Horn, "Artificial Expertise for Process Control," *Mechanical Engineering,* March, 1988, pp. 40–44.
94. Paul A. Lockner and Paul D. Hancock, "Redundancy in Fault Tolerant Systems," *Mechanical Engineering,* May, 1990, pp. 76–83.
95. V. V. Bolotin, *Prediction of Service Life for Machines and Structures,* ASME, New York, NY, 1989, p. 264.
96. N. H. Roberts, *Mathematical Models In Reliability Engineering,* McGraw-Hill, New York, NY, 1964, p. 243.
97. Michael Valenti, "Maintenance Software Keeps Machines Up and Running," *Mechanical Engineering,* November, 1991, pp. 63–65.
98. Jerome Rosen, "Power Plant Diagnostics Go On-Line," *Mechanical Engineering* December, 1989, pp. 38–42.
99. Harris R. Greenberg and Joseph J. Cramer, Ref. 29.
100. Ellen Brandt, "Making Safety Part of the Process," *Mechanical Engineering,* March, 1991, pp. 38–44.
101. John H. West, "The 22-Percent Solution," *Mechanical Engineering,* June, 1984, pp. 52–55.
102. Tony Baer, "Safety at a Glance: Upgrading the Displays in a Nuclear Plant Control Room," *Mechanical Engineering,* January, 1992, pp. 56–60.
103. Henningsen v. Bloomfield Motors, Inc., 32 N.J. 358, 161 A.2d 69 (1960).
104. Escola v. Coca Cola Bottling Co. of Fresno, 24 Cal. 2d 453, 150 P.2d 436, in concurring opinion by Justice Traynor (1944).
105. U.S. Interagency Task Force on Product Liability, *Insurance Report,* U.S. Department of Commerce, 1977. National Technical Information Service, Springfield, VA 22161.

Index

ABOUT THE AUTHOR

Charles E. Witherell is a consulting engineer in private practice who specializes in the diagnosis, analysis, avoidance, and related forensic aspects of engineering failures. He previously worked as principal investigator and project leader at the Lawrence Livermore National Laboratory, and prior to that was vice president of R&D at the Eutectic Corporation. Mr. Witherell holds more than 100 U.S. and foreign patents for his engineering innovations, and has published extensively in technical journals.